LIME

AND OTHER ALTERNATIVE CEMENTS

edited by
NEVILLE HILL, STAFFORD HOLMES
and DAVID MATHER

INTERMEDIATE TECHNOLOGY PUBLICATIONS 1992

Practical Action Publishing Ltd
27a Albert Street, Rugby, CV21 2SG, Warwickshire, UK
www.practicalactionpublishing.org

© Intermediate Technology Publications 1992

First published 1992\Digitised 2013

ISBN 10: 1 85339 178 6
ISBN 13: 9781853391781
ISBN Library Ebook: 9781780442631
Book DOI: http://dx.doi.org/10.3362/9781780442631

All rights reserved. No part of this publication may be reprinted or reproduced or utilized in any form or by any electronic, mechanical, or other means, now known or hereafter invented, including photocopying and recording, or in any information storage or retrieval system, without the written permission of the publishers.

A catalogue record for this book is available from the British Library.

The authors, contributors and/or editors have asserted their rights under the Copyright Designs and Patents Act 1988 to be identified as authors of their respective contributions.

Since 1974, Practical Action Publishing has published and disseminated books and information in support of international development work throughout the world. Practical Action Publishing is a trading name of Practical Action Publishing Ltd (Company Reg. No. 1159018), the wholly owned publishing company of Practical Action. Practical Action Publishing trades only in support of its parent charity objectives and any profits are covenanted back to Practical Action (Charity Reg. No. 247257, Group VAT Registration No. 880 9924 76).

Contents

Preface		ix

SECTION I: INTRODUCTION

1	Alternative Cements: An evaluation of experience ROBIN SPENCE	3
2	Cementitious Materials Projects: Getting the mix right FRANK ALMOND	17
3	Pozzolanic Binders CARL RYDENG	20
4	A View From a Changing Central America: A survival approach for the traditional lime industry CARLOS LOLA	27
5	Manufacture and Use of Lime and Pozzolana Cements in Africa DR A A HAMMOND	35

SECTION II: PRODUCTION OF LIME

Introduction by NEVILLE HILL		47
6	Field-testing of Limestones for Limemaking NEVILLE HILL	49
7	Lime Industry in Sri Lanka J W HERATH	56
8	Locating Reactive Natural Pozzolana W J ALLEN	64
9	Potentialities and Constraints for Using Pozzolanas as Alternative Binders in Kenya RAFAEL TUTS	72
10	Rice Husk Ash Cement R G SMITH	89
11	Research on Development of Alternative Cements Based on Lime Pozzolanas in Uganda for Use in Rural Housing W BALU-TABAARO	105
12	Costa Rica: Small wood-fired kilns JOSE PACHECO	119

13	The Lime Industry in Zimbabwe PHILEMON NHACHI	123
14	Basaltic Rocks for Cement and Other Binders NGO VAN MINH	130
15	Chenkumbi Hills Lime Project JOHN SPIROPOULOS	135
16	Refractory Bricks for Lime Kilns: Small-scale production using local raw materials CHRISTOPHER STEVENS	150
17	Tanzania: Small industries development of lime kilns E G S IKOMBA	156
18	Small-scale Lime Processing: The Balaka experience DRIAN JONES	158
19	Portland-pozzolana Cement from Sugarcane Bagasse Ash SYED FAIZ AHMAD and ZAHEERUDDIN SHAIKH	172
20	KVIC Technology in the Production of Lime and Alternative Cements in India P H NAIK	180
21	Lime Production in Algeria and Prospects for the Development of its Use J MOUSSA	187
22	Technology Options for the Manufacture of Calcined Clay and Pozzolana (Surkhi) J SEN GUPTA	191

SECTION III: USES OF LIME

Introduction by S D HOLMES		199
23	Organic Additives in Brazilian Lime Mortars CYBÈLE CELESTINO SANTIAGO and MÁRIO MENDONCA DE OLIVEIRA	203
24	Lime Stabilized Fly Ash BÜLENT BARADAN	210
25	The Use of Alternative Binders in Rwanda: A case study THEO SCHILDERMAN	218
26	Standards for Building Lime MICHAEL WINGATE	229

27	The Smeaton Project: Factors affecting the properties of lime-based mortars for use in the repair and conservation of historic buildings IAIN McCAIG	237
28	Eurolime: Development and manufacturing of lime for the preservation of monuments JOHN FIDLER	244
29	Lime Stabilized Soil Blocks for Third World Housing DAVID J T WEBB	246
30	Traditional and Current Uses of Lime Mortar, Render and Stucco in Zanzibar FATMA I KARA	257
31	Lime: A common binder for preparation of mortar in earth construction HUGO HOUBEN	262
32	The Landmark Trust: The use of lime-based mortars and plasters in historic buildings projects JOHN BUCKNALL	271
33	Use of Lime: Some techno-social considerations N G DAVE and S K MALHOTRA	277

Appendix 1		289
Appendix 1a *Lime plastering specification*		290
Appendix 1b *Traditional Italian lime stucco recipes*		292
Appendix 2 *The role of the Cementitious Binders Advisory Service (CAS) in disseminating information on lime and alternative cements* OTTO RUSKULIS		296
Appendix 3 *Creating a network: The model and experiences of FAS at SKAT* SUSANNE PREISWERK		302
References		308

Preface

A cementing material or binder is an essential ingredient in most forms of building construction. Portland cement has become the most widely accepted cementitious material, but it is more expensive than other binders and unsuited to small-scale production and, although it is stronger than most alternatives, there are uses for which alternatives such as lime are more suited.

Lime-based renders allow walls to 'breathe', and so allow interstitial moisture to be removed before it can damage the building. Mortars which contain lime are more suitable than OPC mortars for use with earth-based building materials because they are less rigid and more permeable. Lime-pozzolana cements and hydraulic lime can also be used in mass concrete. Lime is versatile, durable, and suitable for use in the most complex and intricate decorative work as well as in simple buildings. The only applications to which these alternatives to Portland cement would be unsuited are reinforced concrete, and high- or medium-strength concretes.

Lime and the many other alternative binders could be re-introduced successfully if more information were available and appropriate building standards enforced. The most successful material in any area depends on local primary resources and the development of appropriate technology. This book brings together expert knowledge from all over the world about the use of different alternative binders, presenting case histories which demonstrate the successes and the pitfalls along with a great deal of practical information based on experience.

A recurring theme in the book is the need to work towards the acceptance of lime, and other alternative binders, by:

- establishing relevant performance standards for alternative cements;
- establishing simple testing procedures to measure compliance with these performance standards;
- holding seminars and training courses;
- gathering comparable case study material on existing production and use of alternative binders;
- initiating research and development projects.

This volume is intended to provide a foundation on which to build a network capable of meeting this challenge.

David Mather
ITDG

SECTION I
Introduction

1. Alternative cements: an evaluation of experience

ROBIN SPENCE*

Technologies for small-scale production of cementitious binders have been under study and development in many parts of the world for the last 20 years. They have been of interest to AT organizations because they possess many characteristics which make them potentially appropriate technologies both for rural development and to provide low-cost housing and infrastructure for the urban poor. They fulfil a universal need; they are small in scale; they have a comparatively low capital cost; they use local raw materials; and they produce for local use.

During this time there have been some successes, but on the whole the diffusion of alternative cements has had limited success, and there have been a number of failures. The reasons for success and failure are complex. Social, cultural and economic as well as technical factors all play a part. The availability, cost and acceptability of alternative materials is a key consideration. Another is the mechanism by which innovation and diffusion is attempted.

In spite of a shortage of reliable data on long-term project histories and performance, it is important to examine and learn from past experience, so that future efforts can be more effectively directed.

This chapter reviews the available information on the worldwide diffusion of alternative cement technologies. It will try to draw some conclusions on the conditions in which alternative cementitious binders can be successfully introduced or promoted; and it will point towards some immediate and longer-term research needs.

Technologies

The different types of alternative cement technology which have been developed can be grouped under four headings:

1. Mini-cement technologies
2. Lime-pozzolana technologies
3. Blended cements
4. Gypsum-based binders

---

*The Martin Centre, Cambridge University, UK.

Mini-cement technologies

Mini-cement technologies take the form of small-scale plants for producing Portland cement, or a binder with properties very close to this. They are based on a vertical shaft kiln, with a daily output varying from 20 to 100 tonnes per day. There has been extensive development work both in India and China, which has taken slightly different forms. In China, the technology was developed as an element of the 'walking on two legs' policy for industrial development: it was assumed that all the cement produced in small-scale plants would be used for low-strength rural applications, and that standards of performance could therefore be relaxed. Very large numbers of small plants have been built, and by the mid-1980s produced over 50 per cent of China's cement[7]. In India, the government insisted that all cement produced in mini-cement plants should meet Indian Standards in full. The technology to achieve this only became available commercially in the 1980s, but it has now proved itself technically and economically viable, and over 100 mini plants are now in operation or under construction[10] with production approaching 10 per cent of India's total output. Although Portland cement is not a subject for technical discussion here, the comparative experience of diffusion of mini-cement with lime-pozzolana technology is worth looking at.

Lime-pozzolana technologies

Lime-pozzolana technologies are traditional technologies for producing low-grade binders in a number of Asian countries, where burnt clay is the usual pozzolanic material used, in the form of either crushed brick or fired clayey soil. The technological developments of the last 20 years have included:

- improvements in the efficiency and quality of small-scale lime production;
- demonstration of the use of a range of pozzolanic materials, including improved types of burnt clay pozzolana, rice husk ash, pulverized fuel ash and volcanic ash;
- the development of new building products using these materials: blocks, roof tiles, etc; and
- the establishment of a number of production plants.

The properties of these binders vary considerably, but they are potentially very adequate replacements for Portland cement (OPC)-based binders in a range of applications. However, mortars and concretes containing them tend to set more slowly and to have a lower final strength than OPC-based mortars. It is impossible to assess the total worldwide production of binders made from these materials, but reports suggest that diffusion of the new technologies has been slow and patchy.

Blended cements

Blended cements are mixtures of Portland cement with either pozzolanic material or with a lime-pozzolana material to achieve a binder with properties somewhere between those of Portland cement and lime-pozzolana mixtures. There have been a number of well-documented projects where such binders have been produced[14,8], but there does not appear to have been much diffusion of these technologies to date.

Gypsum-based binders

Gypsum 'plaster' is also a traditional binder in many areas. Because of its solubility in water its use is restricted to arid areas or internal use, but its strength is adequate for making blocks and roof or wall panels as well as mortars and plasters. In particular locations such as North Africa, there have been a number of projects to develop small-scale production processes appropriate to the local conditions and reports suggest that some diffusion has been achieved.

Analysis of experiences

Most of the technological developments referred to above have been the work of research or appropriate technology organizations. A number of case histories have been described in detail, and an attempt will be made to compare experiences, and to make some general observations about the viability or appropriateness of the technology developed.

The case histories which have been used for this analysis are:

- detailed studies of the development of mini-cement in India and China[9,16]
- reports on lime-pozzolana projects using volcanic ash pozzolana[3,13,19]
- reports on lime-pozzolana projects using rice husk ash (RHA) in India, Pakistan[17] and Thailand[2]
- reports on lime-pozzolana projects using burnt clay pozzolana in India[6]
- report on a blended cement technology project in Botswana[14,22]
- reports on gypsum binder projects in Africa.[18,11]

These case histories all have in common that they describe (in technical and, in most cases, economic terms) projects to establish production plants for the production of cementitious binders at a scale of 20 tonnes per day or less. All are in developing countries. All claim to have developed a binder material which can successfully replace cement in some applications. They are, however, very diverse in other respects: they include a mixture of project descriptions by the organization involved, and evaluations by people independent of the project. Few of them involve a lengthy external study of the project involved. The data given on technical and economic performance are not verified except in the few cases known or visited personally by the author. The use of national currencies at dates which are only approximately known presents a further difficulty of comparison.

Evaluation is thus risky, and only tentative conclusions can be drawn. Nevertheless, it seems important to try to make some assessment at this stage, at least to stimulate some debate, and to try to inform future efforts. There is also an urgent need to establish a programme of research and monitoring of developments in the field of alternative cements to obtain more reliable and consistent data.

Objectives of the projects

The first point of comparison between these case histories is in the stated objectives of the projects. The most commonly expressed objectives (identified as important by more than two-thirds of all projects) were:

- utilization of raw materials
- providing increased employment.

Additional important objectives (of interest to more than one-third of projects) were:

- reducing the cost of cement
- reducing the capital cost of the plant
- overcoming a scarcity of cement
- reducing consumption and cost of fuel
- import substitution.

Less important objectives, but nevertheless of importance to more than one project were:

- reducing transport costs
- training of local people.

It is worth noting that the reports analysed were mostly produced by or for development or AT organizations, and that the stated objectives may be considered as the 'development' objectives, relating to the economic goals of the country as a whole. For those who are interested in managing or investing in industries using such technologies, other objectives such as maximizing income, return on investment, or security of employment might be of more interest.

It is nevertheless valuable to know what these development objectives are because they can be used as a yardstick against which to compare the actual performance of the project. We will examine the most important ones in turn: reduction in cost of cement, local materials utilization, employment creation, capital cost reduction, overcoming cement scarcity, import substitution and reducing fuel consumption. Other criteria of performance will be examined later.

Cost reduction

It is very difficult to make precise calculations to compare the cost of alternative cements with those of OPC. A straightforward comparison of the selling price is not very informative, because generally the products have different performance characteristics and properties. A more meaningful comparison would compare the cost of a unit weight of cement with the cost of the amount of the alternative material required to replace that unit weight of cement in typical applications. The ratio between these two weights has been called the *replacement ratio*.

For mini-cement plants and most blended cements when OPC standard is achieved, the replacement ratio is 1.0. The Indian mini-cement technologies have government certificates to show that their cement meets full Indian Portland cement standards. They sell in the local market at the current price for OPC obtained from normal producers. Three blended cements, however, were sold for between 60 and 75 per cent of the current OPC selling price.

The lime-pozzolana and gypsum technologies differ widely in replacement cost. Indian lime-surkhi appears to offer no cost advantage; Indian lime-RHA, however, costs only 70 per cent of OPC, when equivalent mixtures are compared. Two volcanic ash binders cost 50 and 63 per cent of the replaced cement, with replacement ratios of 1.25 and 2. Indonesian lime-trass blocks are said to be much cheaper than alternative cement blocks but no figures are available. Gypsum plaster produced in one West African project had a selling price roughly 25 per cent of cement, weight for weight, but the replacement ratio is not known.

Thus, while mini-cement technology does not produce a lower-cost cement, the use of the alternative binders, whether blended with cement or using lime or gypsum, does apparently offer substantial cost savings, ranging from 25 to 50 per cent.

Local materials utilization

It is of course a general feature of the cement industry that plants are located near to the sources of the principal raw materials used, limestone and clay (or other siliceous ingredients); thus the objective of local materials utilization is not special to alternative cements or small-scale production technologies. However, conventional large-scale cement plants need very substantial deposits of generally high-grade or uniform materials; the quantity required for 50–60 years operation at 3000 tonnes per day is over 50 million tonnes.

Although there is an abundance of accessible limestone in the world, distribution is uneven and there are regions of the world where occurrences of a size and uniformity suitable for conventional plants do not exist; but such areas often contain small, localized limestone deposits which can be used for some of the technologies described, and the utilization of such deposits has been a feature, if not an objective, of many of the projects.

Two African lime-pozzolana projects are based on small, rather impure deposits of lime; lime for the Indian RHA-lime and lime-surkhi plants and the Indonesian lime-trass plants comes from local small-scale lime producers in areas where cement is not made; and at least one of the Indian mini-cement plants derives its limestone raw materials from sources unsuitable for conventional cement.

Another feature of many of the technologies is the utilization of raw materials for which there is at present no alternative use. Thus the volcanic pozzolanas of Indonesia, Rwanda and Tanzania are extensive and at present hardly used for any other purpose; the same is true of the gypsum sands of West Africa.

Waste materials from other industries are a feature of some of the technologies; rice husk ash, surkhi and pulverized fuel ash (PFA) are in this category. The available quantity of these materials is limited to the supply of the industry concerned, and some of them do have other uses – rice husk as a fuel, for instance, and PFA as an ingredient of Portland-pozzolana cement or as a filler in construction works. Thus there is a danger that what may initially have been seen as a free or waste material could become scarce and expensive as the industry develops.

Thus all of the technologies make use of some unexploited local materials. Many mini-cement plants are located on limestone deposits already used for cement making; and the blended cements make only partial use of local materials, mixed with a proportion of conventionally produced Portland cement. Lime-pozzolana and gypsum technologies are most successful in local materials utilization, but there are potential problems where industrial waste materials of potentially limited supply are used.

Employment creation

Employment creation is generally seen as an important objective of alternative and small-scale cement-making technologies, and most reports gave figures from which employment creation can be estimated. These have been reduced to a common unit of numbers of employees per 1000 tonnes per year production in Table 1 which also gives a comparable figure for a conventional cement plant.

The figures must be treated with caution, as some are derived from number of person-hours' work involved in the production of one tonne during trials; others are based on operating experience, and thus allow for idle time during shut-downs, repairs, etc. They also vary in respect of the actual operations which are included; some do and some do not include raw materials quarrying, for instance. It should also be noted that one tonne of OPC is replaced by more than one tonne of the alternative

Table 1. Capital cost and employment created by some alternative cement projects ('000 tonnes per year)

Technology	Employment (Employees/'000 ty)	Capital cost ($/'000 ty)
OPC plant, India	1.5	120 000–200 000
Mini cement, India	4–16	50 000–80 000
Lime-pozzolana, A	60	7000
Lime-pozzolana, B	15	360 000
Lime-surkhi, India	5	5000
Lime-RHA, India	13	4000–13 000
Blended cement, Africa	16	33 000
Gypsum, Africa	25–40	2500

material in many cases. The employment figures are therefore only indicative.

It is evident from the figures shown in Table 1 that employment per tonne is higher for the small-scale alternative cement-making technologies than for conventional cement plant by a factor of 6 to 20. On the other hand, the total potential employment which can be created by this industry is still comparatively small. As one indication, the use of mini-cement plants in India to create the entire expected expansion of cement-making capacity in that country by 1995 would only add some 100 000 to the working population of a country with over 800 million inhabitants.

Creating employment, numerically, could be regarded as a less important objective than that of creating self-sufficiency and saving capital cost foreign exchange in plant installation. Very high employment may in fact be an indication of inefficient production which could lead to high prices and uncompetitiveness. On the other hand, the experience of industry management and industrial production which can be created by the establishment of numerous small-scale plants in large parent firms, locally managed and independent of government, could have a valuable training function.

Capital cost reduction

Capital cost reduction may be regarded as an objective in all cases even if unstated, and most of the case study reports give information from which capital costs can be estimated. These are shown in Table 1 in which the common unit of capital cost ($ per 1000 tonnes per year) production is used.

These figures are again difficult to compare properly. In all cases given, working capital is included, but in at least one case land and buildings are excluded. Costs have been converted to 1985 by assumed exchange rates which vary with time; and dates are not always available for the figures taken. The capital costs are based on capacity and are not corrected for the effects of capacity utilization, because detailed information is not available. This may significantly detract from the apparent advantages of the smaller-scale technologies where production is intermittent. The replacement ratio of more than one tonne of some materials for one tonne of OPC should also be borne in mind. Finally, substantial differences in the price of the same equipment in different countries adds further difficulties in comparison.

Given these differences, the range of capital cost is still surprising, from a lowest value of less than $2500 per 1000 tonnes per year (almost entirely working capital) in the case of a small African gypsum plant to a highest of $360 000, in the case of a volcanic lime-pozzolana project in Africa. This latter is actually higher than the average cost of establishing a conventional cement plant in Africa and is largely explained by the high import component, and extremely remote location.

Apart from these extremes, the capital costs of the other non-OPC technologies fall in the range of $3000 dollars to $30 000 per 1000 tonnes' annual production, substantially below that of conventional cement plants, for which costs were $120 000 to $200 000 per 1000 tonnes. The cheaper technologies are those in India, where simple machines like ball mills are manufactured locally and are comparatively cheap, or those where only hand tools are used.

The cost of mini-cement plants in India is substantially lower per unit of production than that of conventional cement plants. This is principally because of simpler technology and the possibility of fabrication in small-scale local workshops. However, a very high capacity utilization is assumed, over 90 per cent, which might be difficult to achieve in practice.

Thus, to summarize, capital costs per unit of output can be reduced substantially by using locally fabricated mini-cement plants; and even larger capital cost reductions are possible if blending or lime-pozzolana techniques are used which require only milling and drying equipment, with kilns made from local materials. Gypsum plaster can be made with almost no fixed capital costs if the material is in the form of gypsum sand, and needs no mechanical equipment for processing.

Overcoming cement scarcity

The reasons for the scarcity of cement in many parts of the third world include high capital cost of cement plants, transport difficulties, and foreign exchange costs. In market economies, government policies controlling the selling price from large plants have contributed to scarcity. The existence of an excess of cement-making capacity in the industrialized countries has done little to help developing countries, because they cannot acquire the foreign exchange necessary to purchase large quantities of cement from the industrialized countries. In the long run, the development of indigenous cement-making capacity is, for most countries, the only realistic option.

The small-scale alternative technologies can make a substantial contribution only to the extent that there are adequate raw materials supplies in quantity and distribution for the technology to be duplicated over large numbers of units over a substantial geographical area. To date their contribution has been small, but it could greatly increase in some limited areas where the raw materials are abundant.

Import substitution

The objective of import substitution can be met in different ways, depending on the economic situation. Countries which import Portland cement can achieve substantial foreign exchange savings by manufacturing their own cement, regardless of the scale of technology used. However, these savings are often not as great as might be expected, because conventional large cement plants often require import of fuel or other raw materials, such as gypsum, spare parts and machinery; staff salaries may be repatriated; frequently, too high a proportion of the income generated is repatriated in debt servicing. The total foreign exchange cost of keeping a local cement plant in operation can almost equal the cost of cement in the world market. Thus the establishment of cement-making plants using technologies which minimize imports in capital and running costs can be equally important.

In several Asian developing countries, the cement plant and machinery industry is today totally indigenous, so import substitution is not a consideration in developing local alternatives. In most African countries, conventional cement plants are largely imported and are expected to continue to be so in the near future. Many still depend on oil or imported coal as fuel. By comparison, the four alternative cement case studies from Africa depend to a much smaller extent on imports, as shown in Table 2. The three ingredients of these import costs are capital cost of plant (and debt servicing), transport (both fuel and purchase price of trucks), and the cost of imported cement used as an additive in two processes.

Fuel consumption

In such an energy-intensive industry as the cement industry, energy efficiency is clearly an essential aspect of technology choice. Claims for energy

Table 2. Import requirements of alternative cement-making technologies in Africa

Technology	Approximate % of capital cost imported	Imports	Approximate import content (% of selling price)
Lime-pozzolana A	<5	Transport	15
Lime-pozzolana B	50	Transport, capital costs, cement additive	45
Blended cement	N/A	Cement addition	65
Gypsum	1	Transport	20

savings are made by both large-scale producers (who argue that larger conventional plants are more energy-efficient) and by proponents of small-scale and alternative technologies, because of the different energy requirements of the different technologies.

Table 3 sets out the reported energy requirements of the different technologies described, and the types of fuel used, and compares these with conventional larger-scale cement plants using modern energy-efficient technology.

It will be seen that the fuel consumption of mini-cement plants appears higher than that of modern cement plants by around 50 per cent (though comparable with some of the less efficient plants in operation), while electrical energy consumed is comparable. Lime-pozzolana projects fall into two types. Those using local fuels and burning their own lime have energy requirements per tonne comparable with mini-cement plants; while those purchasing lime and intergrinding it with pozzolana at the plant appear to have lower energy requirements, if plausible assumptions about energy content of the purchased lime are made. The energy content of the small-scale gypsum plaster per tonne is comparable with that of conventional cement plants, reflecting the inefficient calcining technology used.

Given that all lime-pozzolana cements replace less than their weight of cement, it appears that even the most energy-efficient technologies save only a little energy, while the least energy-efficient consume considerably more. However, the cost and import savings from using local fuel (firewood, peat) may well justify the higher consumption of these fuels. Research into ways of saving energy in these technologies is an important further step in their development.

Obstacles to technology diffusion

The primary objectives of alternative cement technologies are to utilize local raw materials, provide increased employment, reduce the capital cost

Table 3. **Fuel and energy consumption of some alternative binder technologies**

	Fuel use t/t binder	Type of fuel	Fuel energy Kcal/kg	Elec. energy kWh/tonne
Large OPC	0.12–0.14	Coal	750–900	110–125
Mini-cement	0.2–0.25	Coal	1300–1625	90–150
Lime-pozz. A	0.37	Firewood	1330	Nil
Lime-pozz. B	0.44	Peat	1452	20
Lime-surkhi, India	0.10	Coal	650	20
RHA-lime, India	0.10	Coal	650	30
Blended cement	0.16	Coal	990	90
Gypsum A	0.25	Firewood	1000	Nil
Gypsum B	0.05	Oil	600–700	Nil

of cement plants and reduce the cost of cement to the consumer. The above evaluation, based on the experience of my case studies using different technologies, has shown that, on the whole, alternative cement technologies can achieve at least some of these development objectives.

Yet all of the case studies have reported numerous difficulties in establishing technically viable and commercially successful production units and thus diffusion of these technologies has been very limited. These difficulties can be grouped into three different categories, each of which will be briefly discussed:

- maintaining the expected or adequate level of production
- achieving and maintaining an acceptable quality of product
- achieving a selling price which is profitable and competitive with existing cement supplies.

Production

In nearly all the case studies described, the technology was either new or untried, and not only the project operators but also the designers of the technology were unfamiliar with it except perhaps at a laboratory or a very small pilot plant stage. Thus the plant project design was carried out on the basis of untested assumptions about production rates, availability of raw materials, etc; proper feasibility studies were not always carried out, and optimistic assumptions were made. This has led (in different cases) to:

- insufficient supplies of raw materials
- raw materials of lower quality, or requiring more processing than expected
- increased transport costs
- increased cost of processing equipment (or lower output of available processing equipment)
- difficulties in recruiting labour/difficulties in recruiting supervising staff with adequate technical qualifications.

Because of underdesign or unforeseen difficulties, funds allocated to the projects initially were inadequate, but negotiations for increased funding were slow and hampered by poor production performance. Production has been frequently interrupted by breakdown of machinery, especially grinding machinery, interruption of electrical power supply and unavailability of essential inputs. These problems are typical of those faced by all industries in developing countries, especially those newly established.

Quality of product

A number of difficulties in relation to quality were reported:

- For mini-cement plants in India, a licence to produce cement for sale

depends on government certification that the cement satisfied Indian standards for OPC. For years, one plant was unable to produce commercially because of failure to satisfy the requirement of that standard with respect to the expansion ratio. Currently, however, all the Indian mini-cement plants are certified OPC producers.
- Standards for lime-pozzolana cements exist in India, and producers of lime-surkhi and lime-RHA cements normally claim to satisfy these standards. Independent investigation indicated that they did so, though the strengths claimed by the manufacturers were in many cases found to be much higher than were actually achieved.
- Outside India, there are no accepted standards for lime-pozzolana mixtures, so quality control standards have to be established by the manufacturers. In two African lime-pozzolana projects, instructions were given to those purchasing the cement on the uses for which it was suitable, and the mix proportions to use. The Indonesian lime-trass industries sell only blocks, which can be assessed according to standards for concrete blocks.
- In the case of one lime-pozzolana project, a single early and prominent case of misuse of the technology had a serious counter-demonstrative effect. In one case cement was initially sold only to selected trained users to guard against misuse.
- The blended cements are claimed to satisfy the strength standards for OPC and to be usable in similar proportions for ordinary concrete work. Independent investigation suggested that this was not the case. Strengths were lower, setting time can be expected to be slower and shrinkage higher than for OPC.
- While individuals, small builders and development projects are prepared to buy alternative cements on the basis of demonstrations of its technical performance and instructions on its use, government departments have been much more cautious. Lack of recognized standards and the possibility of routine testing associated with this appears to be one major obstacle.
- Given the lack of standards, little information is available on which to judge how successful manufacturers of alternative cements have been in maintaining quality. It can be expected, though, that the use of small and variable raw materials deposits, simpler processing technology and labour-intensive methods will result in a much more variable product which needs to be taken account of in devising specifications for use.

Selling price competitive with OPC

- Given the unfamiliarity with the product, lack of acceptable standards and the general conservatism of the building industry with respect to materials, it is unlikely that alternative cements or their products will capture a large market unless there is a significant cost advantage to the

user in doing so. Thus if the replacement cost is higher than about 75–80 per cent of that of using OPC it can be expected that there will be very little take-up of the alternative cements.
- Most of the production problems referred to have an effect on production cost; thus in many cases the production cost actually achieved was considerably higher than originally anticipated at the time of the initial cost calculations. Quality control problems result in generous proportions of alternative cements being specified, or recommended, so that the actual recommended replacement ratio is also higher than could theoretically be achieved. Thus the replacement cost of using the alternative cements is commonly only barely low enough to begin to replace OPC.
- As an essential construction material, price control is often applied to OPC in such a way as to equalize the price the consumer pays in all parts of the country. Such measures can create an artificially low price for OPC, particularly in areas of the country distant from existing cement plants, where transportation costs are high. By keeping the OPC price low, these measures are creating a general disincentive to the development of alternative cement technologies; by equalizing costs throughout the country or region they discourage the establishment of new plants in areas remote from existing cement factories.
- Other forms of government or international agency support to the large-scale cement sector, such as low-interest loans, freedom from import duty on equipment or raw materials, tax concessions on profits, are all forms of subsidy to the large-scale cement sector at the expense of the alternative cement technologies, since without them the selling price of OPC would be higher.

Conclusion

Mini-cement technology is now well-established in India. If offers clear advantages to the producer, and the numbers of productive plants are steadily increasing. Units have already been established in neighbouring countries, and a much wider international diffusion seems likely to follow. But mini-cement technology does not reduce the price of cement to the consumer.

Lime-based technologies for producing alternative cementitious binders (both lime-pozzolana and blended cements) are technically sound, but they have as yet made very limited impact. A number of reasons for this have been identified, the most important of which are:

- the difficulty of establishing a new material in a market dominated by a single well-known product (OPC)
- the fact that the true cost of using these materials, taking account of the need to use a richer mix and their greater variability, may often turn out to be little less than that of the OPC replaced

- most production to date has been in experimental or pilot plants and unforeseen production problems have hampered output
- recognized standards do not exist, and the quality of the binder has often been lower than its proponents claim
- high skills and commitment are required to undertake the complex managerial/technical/financial problems of operating a small cementitious binder plant.

Another difficulty faced by lime-based technologies is that the wide range of different materials which have been developed (using different raw materials and producing to different standards) has not pointed to a single plant design applicable to a wide range of conditions. Given that the most successful pozzolanas are either very localized (such as volcanic ash) or in strictly limited supply (RHA), this seems likely to be a continuing problem.

Nevertheless it seems probable that in areas where there is an adequate supply of a particular pozzolana, a sufficiently concentrated and continuing R&D programme could lead to a replicable small lime-pozzolana plant. This needs to be based on an efficient small-scale lime kiln, and the ITDG kiln developed in Malawi may be suitable. However experience shows that AT organizations are not best placed to take the lead in the commercialization of a technology, and it is essential to identify entrepreneurial interests ready to become involved.

Most of what has been said of lime-based technologies is also true of small-scale gypsum technologies. In this case also the range of types of raw material and possible products is wide, and geographical scope is limited. There is also a more restricted range of uses.

The fact that conditions are not right for the rapid diffusion of these technologies in developing countries is also partly because OPC is still widely available at what are effectively subsidized prices. With dwindling aid budgets and continuing high interest on loans, many developing countries may find that they cannot afford to continue such subsidies as demand for cement grows, and continuing steep increases in the price of cement seem inevitable. This in turn can be expected to create the opportunity for increased use of alternative cement technologies, particularly those which can be established with very small amounts of capital.

AT organizations have a vital role to play in promoting entrepreneurial activity in the alternative cement field and monitoring progress. Some research needs for the immediate future are:

- to set up systematic monitoring of existing production and use
- to identify raw materials resources for known production technologies
- to carry out controlled studies of plant design, both for lime and lime-pozzolana mixtures
- to study user reaction and development of standards for a range of products made with alternative cements

- to study materials in the laboratory and search for new materials which could be the basis of other low-cost binders.

After 20 years of work the field of alternative cements is still new, and beset by many problems and challenges. Yet much has been learnt, and with an adequate infusion of R&D effort, they can be expected to make a steadily growing contribution to building materials supply in the future.

2. Cementitious materials projects: getting the mix right

FRANK ALMOND*

This chapter covers the non-technical factors in programmes of work relating to cements and cementitious materials. It is drawn from, and is illustrated by, examples from ITDG's project work in its building and cementitious material areas, and analyses both success and failure. By their nature, the lessons are reinforced by common experiences in other technical areas, which points up a need to have an evaluation capacity in an organization which can span all of its experience.

It is also a common observation that it is the 'non-technical' factors – economic, social, policy and business – that most often can be blamed for the failure of project work. These non-technical factors are often not wilfully ignored, but are somehow felt to be downstream or second priority issues to 'getting the technology straight'. Even in the most technically-focused R&D activity, these factors should be acknowledged in project design at an early stage.

Objectives

Many problems in this area derive from the absence of clear project objectives, confusing to project members and project partners alike. Objectives should postulate a chain of clear links between project activities and desired outcomes, even though that may be based on a good deal of imperfect knowledge and assumptions at the outset. Gaps in knowledge and assumptions should be clearly stated, and addressed as the work proceeds.

The concept of a 'project' tends to be unsatisfactory in this context, as it encourages a view of a static relationship between project inputs and outputs.

---

*Intermediate Technology Development Group (ITDG), Rugby, UK.

In ITDG, a range of tools and techniques have been developed to assist this process, including:

- Project idea appraisals
- Project frameworks
- Targets and indicators
- Evolutionary work programmes.

Responding to need

Confused objectives are very common when considering how to respond to needs, often based on an imperfect diagnosis of need in the first place. In building materials work, is it the improvement of housing standards for the poor, or income generation, or both, that matters?

Sometimes there is a dislocation between a partner and agent in terms of objectives. It has often happened to ITDG that a partner has a local, social objective which can be pursued in a variety of ways, but our interest has been to prove the validity of a technology for uptake in a wider variety of applications. We do more evaluation and documentation than the partner might feel necessary.

NGOs often have wider campaigning or advocacy roles which should be aligned with partners' views, but often are not.

Products and markets

The choice of appropriate products and their markets and marketing is clearly very closely related to the assessment of needs and potential beneficiaries.

The guiding principle is that the consumers' views should be clearly heard, and their economic/cultural perceptions must be clearly understood.

Economics

Most projects will claim to consider the economics of any technical process. This analysis is often not complete enough, in the sense that the economic performance of a technology has to be assessed not only in its own right, but within the technology or market sector within which it may operate.

At the sectoral level, the fiscal regime within which technologies operate (to the benefit or otherwise of small-scale operators) must be understood.

At the other end of the scale, there is a tendency to analyse micro-enterprises or informal sector enterprises in conventional small business terms, a method of analysis that often does not do them justice or reflect accurately the strategies of small entrepreneurs.

Enterprise

Small entrepreneurs succeed against a fearsome background of difficulties: expensive technologies, lack of skills, no access to cheap finance, scarcity of raw materials, fragmented markets, etc. These difficulties are often not fully understood, and a common temptation is to try to demonstrate the power of new technologies to create new enterprises *at the same time*. Unlike technical support, however, business and enterprise support cannot be contemplated on a short-term, intensive basis. It can, therefore, look expensive, and so local business support is the most attractive option.

Informing

Project work is expensive if there is a heavy donor or assistance input. Sadly, much of the valuable experience that justifies such an expense is left locked up within the project. We are all aware of the duplication of effort that goes on. This underlines the necessity of recording project experience, and the first step is to make sure that enough monitoring is built into the process. The second step is to conduct evaluations, not just to help direct the project, but to understand and synthesize what has been learnt. Then, suitable information can be prepared in the form of case studies, books, training materials.

In ITDG's case, we also see consultancy as a valuable route for extending knowledge and experience. There is increasing potential for networking of 'intelligent' centres for information capture and dissemination. By 'intelligent terminals' I mean the ability not just to distribute information passively, but to enter into an expert discussion with an enquirer, and evaluate the usefulness of information held.

Policy environment

It is common to find that a major obstacle to realizing the full potential of an appropriate technology lies in the national policies that regulate its use (standards and codes most particularly). If so, the project work needs to include work designed to conform to these standards or, where appropriate, to challenge and change them.

The right mix

As well as the above, there are many other factors which people with project experience will be able to list without difficulty. Training, development of local engineering capacity, access to finance, are but a few. None of them are in any way unfamiliar, nor has any radical point been made concerning them. The trick is in getting the mix right; understanding that all these factors are at play to some extent in even the most seemingly simple project. To ignore them, or believe that they can be tackled in due course, is to invite disaster. To try and tackle them all, simultaneously, and head on, is to risk making a project so complex and full of variables as to make disaster certain.

The answer is a realistic appraisal of all of them, and a realistic project strategy that relates them all to the overall project plan, together with a monitoring system that is sensitive to the non-technical as well as the technical performance of the project, and its wider context. In our business, failure is common and expected. Success is all the more welcome when it happens, and chances of success can be improved when judgement is relied on more than luck.

3. Pozzolanic binders

CARL RYDENG*

Building with lime is as old as civilization and numerous building structures give evidence of its durability. Lime in combination with pozzolanic materials was the antique cement, making it possible to build structures standing up to aggressive environments where ordinary concrete of today would fail unless special precautions were taken.

Today lime is not only used as a binder in mortar and plaster, it is also an important input for industry, agriculture and environmental protection. Lime is used in steel refining, in the processing of paper and sugar, in the scrubbing of exhaust gases from power plants and waste incinerators as well as for correcting the acidity of water and soils, as disinfectant and paint (whiting).

Developing countries are therefore expected to develop a lime industry as part of their industrialization process but development circumstances and economy seem to delay the shifting from traditional to industrial processing of lime. The problem is that traditional lime burning requires only modest investment and there seems to be no incentive to shift from traditional to industrial methods as long as only building lime is in demand.

When high quality lime is needed for water purification or some industrial processes, it is frequently imported until the need is so great that the user, for instance a steel or paper plant, sees an advantage in establishing a local production.

---

*Cement Industry Unit, Chemical Industries Branch IO/T, UNIDO, Austria.

In terms of resource conservation, it is increasingly important to shift to industrial processing of lime with minimum fuel consumption to avoid depleting available wood resources with increasing lime production. Less fuel also means less carbon dioxide (CO_2) to the atmosphere and less greenhouse effect for the same quantity of lime.

Another fact of increasing importance is that lime, unlike Portland cement, is not in itself contributing to the greenhouse effect by the release of carbon dioxide since the CO_2 released during calcination is absorbed again during the hardening of the binder.

Finally, it should be noted that a large proportion of the Portland cement used for housing construction could be replaced with lime.

Lime before cement

The early processing of lime was as primitive as an open fire on the field with improvements over the years to accord with local conditions. Sometimes the fire was lowered into a pit with an entry for combustion air at the bottom, at other times the fire was fenced in by a low wall.

The production and use of baked clay bricks led to an increasing need for lime and the establishment of larger kilns. These were built on the ground and, when possible, partly into hills to give easy access to the top and bottom of the kiln to facilitate filling, burning and emptying the installations.

These 'large' kilns were all operated on an average of one firing per month. The operating cycle was started by building up the charge with head-sized blocks at the bottom, starting with an arch to give space for the fire and gradually changing to plum-sized stones on the way to the top. The charging of a five-tonne kiln would require about ten days. The firing of the kiln usually took a week, the cooling three or four days and the emptying another week.

Slaking was mainly done as pit slaking and very often carried out at the building site, since the dough-like slaked lime was heavy to transport, and kept well as long as it was protected from drying out. Lime mortar was, and still is, a very appropriate building material. It is plastic and easy to apply in masonry work and can be kept overnight.

The first Portland clinker was burnt, like lime, in batch-operated shaft kilns. Instead of lime blocks, the charge was built up in dried green bricks shaped with the raw mix known to yield Portland clinker when burned. The higher temperature required for clinker burning led to the use of mixed firing and ultimately to continuously operated shaft kilns which could also be used for lime burning.

Portland cement was for many years no match for lime because of its complicated production process with crushing and mixing raw material, forming bricks as well as the crushing and grinding of the burned clinker into cement which was finally packed and dispatched in wooden barrels. Burned lime was on the contrary easy to handle because it disintegrates into powder by itself when the slaking water is added.

In cement production, crushing and grinding was, in the early days, done with the help of a water-driven hammermill. The finished cement, the fine powder, was separated from the coarse with the help of sieves and the coarse returned to the hammers. Only the invention of the tube mill and the rotary kiln changed the competitive situation between cement and lime.

The technological connection between the lime and the cement industries is demonstrated by the fact that Portland clinker is still produced in shaft kilns up to 200 tonnes per day (t/d) while very large-scale production of lime, for example 1000t/d, is done in rotary kilns. All the merits of Portland cement should not however overshadow industrially produced lime which is environmentally friendly, a stable building material and an important input for industry.

Hydraulic binders

Lime mortars only harden when exposed to the air because the development of strength is based on the recombination of calcium carbonate ($CaCO_3$) from the $Ca(OH)_2$ in the mortar and CO_2 in the air. Hydraulic binders are able to harden and develop strength when immersed in water without access to air.

The most prominent hydraulic binder of today is Portland cement although the Romans used hydraulic lime and Roman cement when special durability and the ability to develop strength was required for marine structures. Examples are the aqueducts built all over the Roman empire from North Africa and Spain to the Middle East, and waterfront structures in at least one antique harbour excavated in the Mediterranean waters.

Other hydraulic binders include mixtures of lime and volcanic ash where names like Santorini and Pozzuoli refer to the areas where volcanic activities produced ash with pozzolanic properties. The name Pozzuoli has given the name used to describe the hydraulic (pozzolanic) properties.

Volcanic ash is a natural pozzolanic material, which only develops hydraulic properties when mixed with slaked lime, in what could conveniently be called pozzolanic mortars, in order to differentiate them from pozzolanic cement, which is normally a pozzolanic binder reinforced with Portland cement. Other important pozzolanic materials are blast furnace slag from steel production and fly ash from power plants. Both occurred as industrial waste with the development of industry and are now extensively used as raw materials and additives for cement and concrete.

A revival of production and use of lime and pozzolanic mortars in the developing countries with the help of present-day technique is technically possible with potential from an environmental viewpoint. Lime-based building methods harmonize well with the requirements of low-rise housing. Small production units are possible with lime and help to secure the supply of a useful binder and at the same time create a seed industry which can breed further industry in the local environment.

The local plant

Depending on the availability of raw materials, the local plant should produce and sell slaked lime (dry), ready-mixed mortars and plasters as well as, when possible, pozzolanic mortars from local pozzolanic materials (fly ash, blast furnace slag or natural pozzolanas). The installations required for such a plant would be: a shaft kiln for lime burning, a small mechanical slaker, crushers and sieves for the grading of limestone (for the kiln), and a mill for the grinding of undersize and pozzolanic materials when available for mixed products. The plant should also include storage silos, transport equipment and a suitable packing machine for the finished products.

The lime kiln could be as small as 5t/d and the mill capacity 2 to 3 tonnes per hour. One important product from the installation should be raw pulverized limestone to be sold as agricultural lime for use by farmers to regulate the acidity of their soils to achieve a good yield.

Besides grinding oversize material for mixed products and agricultural lime, the mill could also be used for the grinding of Portland clinker for special and reinforced products. Finished Portland cement could also be used, but Portland clinker may be preferred since it keeps well but is useless in the local environment without grinding.

Irrespective of the size of the shaft kiln, it should be a tall slim industrial construction with well-defined preheating, burning and cooling zones to give good fuel economy. It should have external firing boxes or burners to avoid impurities from the fuel in the lime. It should be easy to feed and operate in a controlled manner at different levels of production, within certain limits, depending on demand. Fuel consumption should compare well with large-scale industrial kilns.

In order to reduce investment costs in the described small kiln, it is presumably necessary to feed raw stone and extract burned lime manually, in controlled quantities. Such a plant should, for economic reasons, progress in small steps where each development builds on the previous one so the available capacity is fully utilized at all times.

A complete small-scale plant, as described above, is to our knowledge not yet established anywhere and one of the reasons for presenting such a plant idea is to encourage the establishment of such plants to prove their viability and value for rural development.

Development phase I

The first development phase should focus on the establishment of the proposed type of kiln with simple slaking and packing facilities. Undersize material would in this phase be piled up for later use if and when a mill installation would be available. Raw lime would be quarried as simply as possible and processed into quicklime and slaked lime for sale to builders with support and advice to the users on how to apply the binders efficiently.

Development phase II

The second development phase should focus on adding a mill installation, storage and mixing facilities so agricultural lime and mixed products, such as ready-mixed mortars, could be produced and sold. At this point, the material formerly rejected as undersize would, depending on moisture content, be sold alongside the hydrated lime.

Development phase III

The next phase is establishing test facilities and the integration of pozzolanic materials or Portland clinker in the processing of everyday building materials suited for local housing construction. In this connection, it should be considered that multi-storey buildings constructed with lime mortar were common in Europe and the US before Portland cement was available.

Preparatory studies

Small plants are either built up with the available means and developed into increasing sophistication, if possible with the help of their own earnings, or promoted through preparatory studies aiming at mobilizing the necessary resources and loans required, at least, for the initial investment.

The classical preparatory study is meant to advise the entrepreneur and his bank about the feasibility of the development idea and help to mobilize loans or to prove that the envisaged investment is not attractive.

A feasibility study organized by UNIDO would first of all check market conditions for the intended products and continue with a survey of the availability of raw materials. Finally, realistic estimates for investment and operating costs would be used to calculate the economics of the proposed venture inluding information on the expected sales price, pay back period and return on the investment.

The situation today is that no major investment is made without a feasibility study which in reality serves to promote investments or to warn against doubtful ventures. In this connection it should be considered that it is better to lose US$50 000 in a negative feasibility study than to throw away US$5 million in a loss-making investment.

Case study in the development of pozzolanic cement production

In 1984, UNIDO was invited by UNDP to be technical support agency for a pozzolanic cement project started one year earlier by a United Nations volunteer. Taking part in the project was easy because the initial work had already been done by the UN volunteer while bilateral funding had financed some equipment. UNIDO actually inherited a small and primitive batch mill which had been commissioned a few months earlier along with the volunteer who had designed the mill.

The initial recipe for the pozzolanic cement was 5 parts of boiler slag (ash), 4 parts of rejects from a small brick plant (activated clay) and 4 parts of hydrated lime. The mixture was pre-ground to a cement-like fineness and finally mixed and ground with equal amounts of Portland cement.

Since UNIDO was inheriting the established production of a binder, of which 50 per cent was Portland cement, we saw no performance risks in the product and found it logical to use the facilities to examine limits and possibilities for the production of a pozzolanic binder with locally available materials.

Our first idea was to propose a small laboratory established near the production facilities so the quality of the inputs could be controlled together with the finished product(s). The idea was to engage in applied research to examine how the recipe could be safely modified to include less Portland cement which was the most expensive raw material in the pozzolanic binder.

The existing equipment also had some shortcomings so part of our technical assistance activities focused on improving the installations and working conditions in order to improve efficiency and reduce possible health risks from the dust-laden work environment.

All our good intentions were however blunted by shortage of funds and the idea of building up laboratory facilities and at the same time improving and operating the existing installations proved impossible in the short term. An integrated proposal for the development of local building materials production in the region, including the described assistance to the pozzolanic cement plant, was rejected and we ended up with a piecemeal approach focusing first of all on operating the plant and only at the very end making it possible to build a simple laboratory.

By the time we had completed the laboratory and were ready for process control and research, a major accident with the mill stopped the work. When the mill was repaired, the counterparts forgot to lubricate the transmission and the new gear wheel was as worn out as the old in no time. When finally both the production equipment and the laboratory were ready for the planned research work, the counterparts ran out of funds and all UNDP-sponsored UNIDO activities to the project came to an end.

Looking back, we came to the conclusion that it would have been better to engage only in the applied research activities with the support of an experienced consulting company or a cement factory. The locally available materials should have been examined in a well-equipped laboratory and a cash flow analysis worked out before any investment decisions were made.

At the time of the request for assistance we only saw an ongoing production with certain problems and the need for help and we had no idea that our efforts would be hindered by a piecemeal approach draining the efficiency of our work. We only saw a choice between complete, well-organized assistance and no assistance at all. Today it is difficult to judge whether a job half-done was better than nothing.

Promotion of pozzolanic binders

Pozzolanic materials, whether natural or artificial, represent a value related to the properties they can add to a binder, and the savings in fuel and raw materials they can help to realize. Not all pozzolanic binders are, however, equally well accepted by potential users because they are accustomed to using Portland cement and anything less makes them sceptical.

Before Portland cement was invented and made popular, only building lime was generally available. Other binders, slightly better in terms of strength and performance, such as Roman cement, hydraulic lime and lime-pozzolanic mixtures, were welcomed because of their improved performance under certain conditions.

Today, when Portland cement has reached the most remote corners of the globe, nobody seems really to need the products just mentioned, but everybody speaks about pozzolanic cement expecting a low-cost product. In our opinion, the main point today is not low-cost cement, but resource conservation aimed at giving coming generations similar development conditions to what we have.

The most efficient way to conserve energy in cement production is to add normal Portland cement to pozzolanic materials while the general properties of the cement remain unchanged for normal uses. Attempts to market lime-pozzolanic mixtures without reinforcing Portland cement are unwise because the concrete is water sensitive. Incorrect applications spread a negative reputation, making alternative binders practically impossible to sell.

The other possibility is to make improved lime binders for use in mortar and plaster as well as soil stabilization (walls) and road foundations. The problem here is that it is not enough to develop a product and design a method for its manufacture, which is all fairly easy. The main obstacle is to teach users how to apply the 'new' products so they perform well and open markets without major problems.

The latest similar experience was the drive to make rice husk ash popular. Today not a single rice husk plant is reported to be in operation.

Another effort was the promotion of small-scale shaft kilns in India for cement production. General production problems and quality variations apparently gave the upper hand to large-scale modern plants now built all over India by private owners. An important incentive for this development was deregulation of the cement price to encourage more competitive market conditions.

China is still building shaft-kiln cement plants in remote locations but an all-out initiative to promote large-scale modern plants is expected to overshadow the domestic shaft-kiln installations in a few years and make such installations obsolete. China is already engaged in the export of modern and large-scale cement plants to other developing countries and the need for resource conservation at home and abroad makes the modern computer-controlled factory a front runner.

The best small-scale plant today is a simple lime factory matching local needs and able to deliver building lime and chemical lime at prices comparable to Portland cement. Such plants are needed mainly in Africa to spread development more evenly over the waste territories of the African states.

State control and government regulation have, however, made the investment climate unattractive to small entrepreneurs. The result has been that only the government and a few large international companies have invested in large-scale installations.

Experience indicates that we speak about building small factories and we do it long enough to arrive at the need for large-scale installations which are then built. Examples include Benin (former Dahomey), Congo, Gabon and Togo. Small-scale plants were actually built in Cameroon, Mali and Niger but they all met problems in continuing and efficient operation.

4. A view from a changing Central America: a survival approach for the traditional lime industry

CARLOS R LOLA*

The lime subsector is an important component of industrial development and economic growth of Central American nations. It not only provides valuable inputs to the manufacturing, agriculture and construction industries, but could be a foreign exchange earner, as well as providing a source of employment.

This chapter describes the efforts of AT International (ATI) to upgrade the traditional lime industry in Central America. The first section outlines

---

*AT International, Washington D.C., USA.

the region's socio-economic conditions, while the second reports on the traditional lime industry in the seven countries of Central America. A description of the regional programme design follows, outlining how it has evolved from a pilot project in Costa Rica. The chapter concludes with a summary discussion of the need for a survival strategy for the Central American lime industry.

The Central American isthmus, a sub-region of Latin America, has approximately 30 million people located in seven countries: Belize, Guatemala, El Salvador, Honduras, Nicaragua, Costa Rica, and Panama. The average per capita GNP for the region is approximately $1210 in 1989 dollars[1].

The decade of the 1980s in Latin America (population 431.9 million[2]) is considered the 'lost decade', characterized by economic stagnation, a debt crisis, political upheaval, and degradation of social conditions in most countries. Today, most of Latin America is ruled by democratically elected governments that encourage free-market economic strategies for development. Re-evaluation of past economic policies has been underway, with most countries adopting economic stabilization and structural adjustment programmes to establish a more solid foundation for future economic growth. In addition, most countries have been, or are considering, lowering trade barriers to stimulate development and competitiveness of domestic industries. Furthermore, a revival of sub-regional integration initiatives in Latin America makes the future economic outlook more optimistic. Among the most notable sub-regional initiatives are the following:

- North America Free-Trade Area
- Central American Common Market
- The Andean Group
- The Caribbean Common Market
- The Mercosur.

My purpose here, however, is not to analyse current economic events in Latin America, but to outline the current socio-economic context in which the traditional lime industry in Central America takes place.

ATI experience

ATI's knowledge of small-scale lime processing is the result of extensive research, information gathering, technical assessments, and direct field work with local organizations in Botswana (1984–87), Sri Lanka (1987), and Costa Rica (1984–91).

In Central America, ATI has been working over the past six years with the Technological Institute of Costa Rica (ITCR) to identify and address bottlenecks to production in the traditional lime industry, identifying potential for technological changes and enterprise development strategies. A modified version of a traditional kiln has been developed, with the following features:

- 40–45 per cent higher production output,
- fuel consumption 10 per cent less than the traditional one,
- reduced exposure of lime workers to high temperatures and excessive lime dust, and
- a lower investment cost.[3]

ATI also identified modern processing techniques available in Spain, India, and Germany.

In July 1991 in Costa Rica, ITCR, ATI and the United Nations Center for Human Settlements (UNCHS) co-sponsored a Regional Workshop on Small-Scale Lime Production in Central America. The workshop brought together over 40 representatives of the lime subsector from Belize, Guatemala, El Salvador, Honduras, Nicaragua, Costa Rica, and Panama. Participants included small producers, government officials, representatives from development banks, research institutions, and international development/donor agencies.

Lime industry profiles prepared by the participants of each country provided a preliminary picture of the lime subsector in the region. In addition, the participants organized a Regional Lime Group to plan, promote and co-ordinate activities in the lime subsector.

The traditional lime industry in Central America

Limestone is a mineral resource abundant throughout the Central American region. Small-scale lime processing is characterized by low productivity and low quality output. As a result, producers in the region have not been able to meet the demand for lime and lime products. Small-scale producers suffer from limited marketing capabilities and lack of entrepreneurial support.[4]

In the early 1500s, the Spanish introduced vertical shaft kilns and indigenous production techniques that have changed little over the years and are still in use. Although variations in the level of development exist among countries of the region, most small-scale lime producers experience similar problems. These include: low prices for their product, limited access to profitable markets due to poor quality, limited access to working/investment capital, high firewood consumption, low productivity, exposure of workers to high temperatures and excessive lime dust during kiln firing and unloading. There are approximately 600–800 producers of lime in the region, producing roughly 350 000–450 000 tonnes of lime per year with an estimated value of US$17.6 million.[5]

Although there are a few large lime production facilities in the formal sector, small- and medium-scale lime production accounts for over 80 per cent of the total output. With few exceptions, lime production facilities are owned and operated by independent entrepreneurs often organized in small producer associations. These small and medium enterprises, usually found in the informal sector, are not receiving assistance in the areas of enterprise development, access to technology alternatives and financial resources and credit.

Studies of the Costa Rican lime industry by ATI and ITCR indicate that the small-scale lime industry suffers from poor management, lack of an effective marketing strategy, poor quality product and poor quality packaging. The industry also has a high firewood consumption rate.

The studies also indicate that low profit margins were caused by low prices assigned to poor quality output, or what consumers perceive to be poor quality products. For example, hydrated lime that is milled usually sells for twice the price of its counterpart of similar chemical composition without milling, yet consumers are erroneously judging quality by the product's appearance. The study also revealed revenue differences between quicklime and hydrated lime producers, attributed to lack of adequate management and production organization.

Small-scale producers primarily employ traditional methods of extracting and processing limestone. This has contributed to the decline of the industry in recent years. Quarrying is done often with a chisel and mallet. The lime kilns are characterized by low thermal efficiency and uneven thermal distribution resulting in tremendous heat loss and irregularly calcined quicklime. Hydration is often carried out manually and hydrated lime is of poor quality due to contamination with ash and sand and because it is packaged in used sacks.

Costa Rica produced approximately 16 725 tonnes of lime in 1988[6] all of which was locally consumed. The potential market for lime products is estimated to be up to 62 215 tonnes per year, if alternative lime uses are promoted and quality is increased.

While many of the findings of the Costa Rican study can be applied to the Central American region, each country presents its own set of challenges and opportunities for the lime industry. A study of the Guatemalan lime industry[7] concluded that while much of the production is at the small and medium scale, these producers face bottlenecks due to lack of financial and technical assistance. Lime production in Guatemala represents approximately 50 per cent of Central America's total production. Guatemalan demand for hydrated lime was estimated at 154 261 tonnes in 1988 and is increasing by approximately 6.4 per cent per year. The price paid for hydrated lime has also steadily increased to between 18 per cent and 42 per cent per annum. However 66 per cent of hydrated lime production is controlled by three producers. Small and medium producers find it difficult to compete as they do not have access to adequate hydration equipment. In 1989, 314 tonnes of hydrated lime were imported into the country. The principal markets for hydrated lime are in industry, agriculture, and construction.

In Guatemala, small producers concentrate on quicklime production. Therefore, these producers do not have access to markets that pay higher prices for their products. The principal markets for quicklime are in industry, construction and tortilla making. Because of crude extraction methods, the limestone that is extracted is very irregular. Guatemala also lacks information on the location and extent of limestone deposits in the country. In Guatemala, small producers are facing increased prices for scarce firewood for kiln fuel and are thus looking at alternative fuel sources such as sawdust and coffee bean husks.

Estimated demand in El Salvador[8] for 1989 was 24 620 tonnes. Demand is growing by approximately 5 per cent a year. Lime is used primarily in construction, sugar refining, agriculture, water treatment, and leather tanning. Preliminary estimates show that approximately 25 per cent of the hydrated lime consumed in the country is imported.

The kiln design used in El Salvador artisanal lime production is a rustic one, fashioned from a simple adobe-reinforced cylinder with a thermal efficiency of approximately 22 per cent. The severe deforestation in El Salvador has forced lime producers to use alternative fuel sources without regard to the environment. For example, used rubber tyres are being burned, releasing noxious sulphur gases into the atmosphere which leads to acid rain.

In Nicaragua, approximately 12 750 tonnes of lime are produced annually.[9] Approximately 80 per cent of hydrated lime consumed is used in the sugar industry. Calcium carbonate is used as an input in the production of soaps, paints, and talcum powder. Limestone deposits are in abundance in the north, south-west and eastern (Atlantic coast) regions of the country with the greatest deposits being on the Atlantic coast.

Currently most small-scale processing is conducted in the south-western part of the country. Small-scale production is characterized by vertical shaft kilns fuelled by firewood. Extraction of the limestone is done manually. As with many countries in the region, Nicaragua faces increasing deforestation and shortages of firewood. Current production methods are not economically viable and cannot be sustained.

A study of the Nicaraguan lime sector concluded that the introduction of improved technologies could greatly increase production for small-scale producers. Improving the technical efficiency of production, as well as exploring alternative fuel sources are the two keys to the survival and growth of the lime sector.

Honduras produces approximately 43 200 tonnes of lime per year[10]. While exact statistics on consumption are unavailable, it is common knowledge that the entire national production is consumed locally and lime is being imported. Lime is used mainly in water treatment facilities, as well as in the preparation of beverages and paint production.

There is a demand for a higher quality product which local producers are as yet unable to offer. The national lime industry suffers from lack of quality control standards and poor technologies. Antiquated production technologies are partly to blame for the low quality product. Quicklime and hydrated lime are often contaminated with ash and dirt, making it difficult for the products to compete in markets demanding high quality lime. There are few facilities to identify the chemical composition of the limestone; therefore, improved quality controls cannot be implemented.

The typical artisanal lime kilns in Honduras are of a traditional design which requires manual discharge after calcination. This results in delays between batches, exposing workers to extremely high temperatures. The time to complete a batch of quicklime including loading the kiln, firing time, cooling off, discharge, milling and packaging can run between 10 and 15 days. Thus, at a maximum two burnings can take as long as a month, while in other Central American countries small producers have the capacity to perform up to four firings per month. There are approximately 300 artisanal kilns in Honduras, but in order for this small-scale industry to survive and grow it will need to upgrade its productivity.

In Belize the local market demand for lime was 55 663 tonnes in 1990. Of this, approximately 80 per cent was supplied through imports from Mexico.[11] While adequate lime deposits exist in the country, little attention has been paid to developing this industry. Small-scale lime production may be suitable as a small enterprise activity for refugees from El Salvador and Guatemala.

Proposal for a regional programme for small-scale production of lime in Central America

As a follow-up to the regional workshop in Costa Rica in July 1991, ATI, in collaboration with ITCR, has prepared the following proposal.

The overall development objective of the regional programme is to increase the incomes of small producers and workers in the traditional lime industry through higher productivity and higher product quality. A further objective is to foster the development of the industry in an environmentally sustainable way[5]. Since the characteristics of small-scale lime production are so similar throughout Central America, this programme is particularly conducive to a regional approach.

The programme has two main components:

- Technical co-operation
- Investment financing

The first component will involve subsector mapping of the lime industries. These analyses include mapping the backward and forward linkages, identifying principal functions, participants and channels through which raw materials are obtained, transformed and marketed to final consumers. Following these analyses, feasibility assessments for the targeted areas will be conducted including economic and commercial analysis as well as studies of the impact on women and the impact on the environment. Based on the outcomes of the feasibility assessments and impact studies, project interventions will proceed.

Technical assistance will be provided for both the engineering and institutional aspects of project design, implementation and evaluation. The project staff will work with small-scale producers, co-operatives and associations to improve business management and organizational capacities. Potential project interventions include: improved quarry management, introduction of energy-efficient kilns, quality control standards, efficient hydration methods, and improved product packaging. The proposal considers the upgrading of traditional lime processing techniques as the first level of intervention. This approach is frequently the most effective way of increasing the productivity of small enterprises, particularly those owned and operated by the rural and peri-urban poor.[12]

The second component will consist of a capital investment fund to finance improved lime processing methods and production practices, as well as working capital. A selected group of institutions will manage the lime investment fund. In addition existing funds for small enterprise programmes will be explored for use by lime producers in the region.

The project will be co-ordinated by a Regional Project Management Unit (PMU). The PMU will most likely be located in Costa Rica and will provide overall co-ordination to the project as well as technical and managerial support. The PMU will consist of Chief Technical Advisor, a Technical Officer, and a Business Development Specialist. In each of the five countries a local implementing organization will be selected for project management and implementation. This local organization should have experience in carrying out international development projects, working with small-scale entrepreneurs and familiarity with the lime subsector. The local implementing organization may be a research institute or a non-governmental development organization.

The PMU and local implementing organizations will work closely with lime producers either through co-operatives or independent small entrepreneurs. The active participation of the lime producers will be sought in all phases of project design and implementation.

Project impacts: Among the expected benefits of this regional programme are the following:

- increased demand for domestically produced lime, due to better production
- an increase by 30 to 40 per cent in the income of the small producers and workers of the lime industry

- an increase in the productivity of local businesses through access to improved small-scale lime processing technologies and business development skills. An estimated 600 to 800 lime producers could directly benefit
- amelioration of adverse impacts on the environment by reducing firewood consumption through fuel-efficient kilns and controlling air pollution through air filters
- generation of foreign exchange savings through domestic production of high quality material to reduce reliance on imported lime and other substitutes
- stimulation of other sectors of the economy. Agriculture, manufacturing and construction will all benefit through the increased availability of lower-cost higher quality lime products.

Survival strategy

The traditional lime industry in Central America needs to recognize the changing economic environment, and that industrial competitiveness is the critical element for its survival. Ironically, at the beginning of the 20th century, the US lime industry was much like the lime industries of Central America today. Mass production and 'hi-tech' transformed the US industry drastically but changed little or nothing in Central America and other small countries.[4] Now, in the 1990s, the small-scale lime producers of Central America need to implement a strategy to pursue business success and sustainability in the medium and long term. This is important not only because of the direct benefits to small producers in increased productivity and incomes, but also because, strategically, small producers in the aggregate can make a larger contribution to sound economic development more rapidly than additional costly capital intensive investments. This strategy is outlined here in the form of a proposal for a regional programme.

In recognition of this programme's potential, representatives from the Economic Ministries of the region, the Central Economic Bank for Economic Integration, several producers co-operatives, and the United Nations Program for Especial Assistance to Central America have expressed their interest in supporting and participating in this regional initiative.

Pilot projects similar to the one supported by ATI in Costa Rica, as well as the proposal described here for the Central American countries, are potentially replicable in other areas of the world. These initiatives could be useful tools to optimize use of mineral resources, to develop inter-regional trade and co-operation, and develop competitive regional lime industries. The future of small-scale lime producers relies on their access to appropriate technologies, financing, profitable markets, and use of sound business practices.

5. Manufacture and use of lime and pozzolana cements in Africa

DR A A HAMMOND*

Ordinary Portland cement is a strategic material in the construction industry. It is expensive in many African countries and requires foreign exchange. Foreign exchange shortages in most African countries very often also lead to corresponding shortages of ordinary Portland cement. These factors constitute a major constraint in any attempt to provide shelter in most African countries. The availability of cement therefore determines the rate at which construction of building or other infrastructural facilities are completed. Because of this, it is not uncommon to see half-completed structures abandoned due to lack of building materials, especially cement.

Demand for cementitious materials

Cementitious materials comprise Portland cements, lime and lime-pozzolana or cement-pozzolana. The need for cementitious materials in the developing African economy is a reflection of the measured need for houses, schools, factories, offices, roads, and communication as well as other facilities, and of the rising quality standards and users' requirements for such buildings and facilities. Since Africa is a developing continent she will necessarily need large quantities of cementitious materials for her construction programmes for development. In many construction processes ordinary Portland cement can be substituted to advantage by lime or pozzolana cements.

Prospects for developing lime and pozzolana cements in Africa

The prospects for increased production and use of lime and pozzolana cements in Africa are great and should be viewed in terms of availability of resources: raw materials, technology, labour and capital. The African continent is rich in raw material resources for building materials production. The continent has all the necessary raw materials it needs for producing lime and pozzolana cements. In addition, fuel and electrical power as well as water are available.

---

*Building & Road Research Institute (CSIR), Ghana.

Lime production

Lime was the main cementitious material for construction in Africa, specifically, in Egypt and many Northern African countries, many centuries before the manufacture and use of Portland cement were introduced to the region. Practically, every country in Africa has occurrences of calcium carbonate in the form of limestone, calcite, calcrete, sea-shells, clam shells and dolomite which form the basic raw material for lime as well as Portland cement production. In some African countries such as Algeria, Egypt, Ethiopia, Zambia, Tunisia, Togo, Zaire, Nigeria, Morocco, Mauritania and Cape Verde, the limestone deposits are vast and could support large production ventures for many years. Other countries have small deposits in scattered locations across the country. In Somalia and other countries situated along the eastern coast of Africa, Tanzania, for example, there are extensive deposits of coral limestone. These deposits are generally shallow and occur between layers of sand. It is therefore easy to quarry manually. Lime production on a small scale is a flourishing industry in Mogadishu and other coastal towns in Somalia. Other forms of calcium carbonate occur in many parts of the continent. In Botswana, for example, there is a vast deposit of calcrete at Segeng some 50km west of Kange in the south western region of Botswana.

Another incentive to develop lime for construction purposes is that the great demand for Portland cement cannot be satisfied by most African countries. And so, although lime has a slower setting time than cement, and the final strength is lower, it is perfectly adequate for most low-strength applications, such as mortars, plasters, and as a stabilizing agent for improved soil blocks. Lime can be a perfect substitute for such applications because it has, in addition, good workability, ability to accept movement without cracking, water retentivity and resistance to water penetration. It has also a better endogenous healing property in stabilized soil blocks. Used with a pozzolana, the lime pozzolana can be substituted for cement in masonry work where the high early strength of cement is not required.

Pozzolana production

As Portland cement is virtually the only cementitious product used in Africa for construction, with a high foreign exchange content, it is obvious that any local cement extender will bring great savings in foreign exchange. Pozzolana is such a material. A pozzolana is a siliceous, aluminous or ferrogenous material, which by itself is not cementitious, but which under a certain state of crystallinity and structure could react with lime in the presence of moisture at ordinary temperature and pressure to yield cementitious products. By this definition, pozzolanic activity, therefore, depends on the fixative property of the material in respect of calcium hydroxide and its ability to harden under water as a consequence of changes that the above reaction produces. These characteristics may be separate, and it may happen that, while large quantities of lime are fixed by the materials having pozzolanic activity, the accompanying cementitious properties are quite moderate. Malquori[1] suggested in 1960 that materials having pozzolanaic activity should be defined, in respect of their use, as cementitious materials, quite apart from the interpretation of the chemical and physico-chemical phenomena which are responsible for the hardening of the hydraulic binder.

Pozzolanas are grouped into two main classes, namely, natural and artificial. Though natural and artificial materials have similar pozzolanic activity, they differ greatly from one another in origin, but slightly in chemical and mineralogical constituents.

Natural pozzolanas

Natural pozzolanas may further be divided into two main groups:

- There are those derived from volcanic rocks in which the amorphous constituent is glass produced by fusion. These are, for example, volcanic ashes and tuffs, pumice, scoria and obsidian.
- The others are derived from rocks or earth for which the silica constituent contains opal, either from precipitation of silica from solution or from the remains of organisms. Examples of these are diatomaceous earths, cherts, opaline silica, lava containing substantial amounts of glassy component and clay which has been naturally calcined by heat from a flowing lava. In Africa, some of the known sources of natural pozzolanas of volcanic origins may be found in Cameroon, Cape Verde, Burundi, Ethiopia, Tanzania, Kenya, Rwanda and Algeria.

Artificial pozzolanas

Artificial pozzolanas may also be divided into two main groups, those of inorganic and those of organic origin. The most important artificial pozzolanas of inorganic origin are obtained from calcined clays and shales, calcined bauxite, calcined bauxite-waste, calcined spent oil, calcined moler, calcined gaize, 'fly ash' (pulverized fuel coal) and surkhi (brick powder).

The sources of artificial pozzolanas of organic origin are ashes of rice husk, coffee hulls, coconut shell, sugarcane bagasses and palm-nut shells and fibres. There are investigations into the use of cocoa pod for pozzolana production. Of these pozzolanas of organic origin, rice husk ash has been well investigated and documented.

When rice husk is burnt at appropriate temperatures, the ash that comes out of the burning is predominantly active silica which can react with lime to form cementitious products. Practically all African countries grow rice and therefore the prospects of using this agro-waste for pozzolana production in Africa are bright. As agricultural waste, it has no commercial value and poses a problem of disposal. Since these agricultural wastes are fairly well distributed in farming areas which are essentially rural and in suitable quantities, it is possible to establish small-scale industries for their exploitation.[3]

Fly ash is widely used in Europe, the US, China and India under numerous commercial names. In African countries where coal is used for industrial purposes, e.g. Botswana, Nigeria and Zimbabwe, the ash can profitably be used as pozzolana. Practically, there are clays everywhere in Africa which can be used if they are found to be suitable. The oil-spent shales in Nigeria, Gabon, Libya, Algeria and Egypt can also be used in producing pozzolana.

Calcined bauxite is known to make an excellent pozzolana but it is more profitable to use it for the manufacture of aluminium, particularly in countries where clays and shales, fly ash and rice husk are easily available as substitutes. In Ghana, extensive work has been carried out on conversion of bauxite-waste to pozzolana.[2]

Energy for lime and pozzolana production

The prospects for pozzolana and lime production in Africa are good in terms of the energy required to convert the material into pozzolana if it is of an artificial type. Energy requirements for manufacturing a kilogram of Portland cement clinker is considerably higher than for lime or pozzolana. The energy consumption for the former ranges from 1220 to 1440Kcal/kg. Energy requirements are higher for the wet process than for the dry one. Favourable energy consumptions are obtained with modern rotary kilns which are equipped with special devices for achieving complete utilization of all available source of heat. Again, energy requirements are higher for the wet process than for the dry one. Energy requirement for manufacturing a kilogram of clinker can be as low as 840Kcal for such kilns. But, for pozzolana production this type of technology is not required. Table 1 shows energy consumed during the manufacture of lime, pozzolana and ordinary Portland cement. It is clear from the table that much more energy is required for producing clinker even if energy consumed in grinding is not taken into consideration. This means that pozzolana can be produced at relatively low cost. There is therefore some cost-saving to be derived in using cements which have been locally produced for housing.

Technology

This section reviews the technologies for producing lime and pozzolana cements. Unlike many building materials such as glass and steel which are produced by high technology, the technologies for producing lime,

Table 1. Energy inputs of materials[3]

Materials	Solid fuel Kcal/kg	Electrical energy Kcal/kg	Total energy Kcal/kg
Lime	722	3	727
Hydrated lime	547	50	597
Fly ash	0	0	0
Surkhi (brick powder)	0	50	50
Burnt clay pozzolana	228	56	284
Rice husk ash	0	50	50
Ordinary Portland cement	1440	120	1560

pozzolana and bricks and tiles are available at different levels of mechanization. Broadly, these methods may be grouped under labour-intensive, semi-mechanized and fully mechanized methods. All of these levels of technology are available for adoption or for adaptation.

Labour-intensive technologies

In many African countries, labour-intensive methods exist for producing lime and pozzolana on small-scale industry basis. In Somalia, for example, where there is a flourishing small-scale lime industry, two simple technologies are employed. These are clamp firing and shaft kiln firing. In the clamp firing, firewood is used. This is stacked in layers with coral limestone to form a clamp with provision of flues for gases to escape. This process takes 15 days to complete the calcining. There is no control whatsoever on the process. The burnt lime is exposed to the atmosphere and it is hydrated by sprinkling of water. The amount of water required for hydration is gauged by experience. Similar simple methods are employed throughout the African continent. Production levels vary considerably from 0.5 to 2 tonnes per day for small-scale production. This technology is labour-intensive. Simple equipment such as wheelbarrows, rake and a water drum for storing water are all that is needed. The beauty of it is that it lends itself or adaptation and renovation.

Simple labour-intensive methods are also used for producing pozzolana. In most places where the pozzolana occurs naturally, the material is won with pick-axes and shovels and is immediately used for block or mortar-making without further processing. This is the most common practice in Tanzania and Rwanda, and also in Indonesia.

Where the pozzolana is produced artificially as from clays and agricultural waste, simple labour-intensive technologies are employed. In the case of surkhi, the clay is fired with wood, coal or rice husk and the calcined material is ground.

Usually the material is ground using 'animal power'. In Malawi, a renovation was made using a corn mill to grind the pozzolana. Pozzolana derived from rice husk ash is also simply obtained by burning the rice husk waste for the ash using simply designed kilns. Since there is an optimum temperature for calcining both the clay and the rice husks and since the methods used generally ignore this fact, pozzolana manufactured from these methods is normally of poor quality.

Semi-mechanized technologies

The semi-mechanized technology allows for various degrees of mechanization or combination of labour and capital. The choice of relative levels of labour and capital depends on many factors including the required production scale and availability of labour and capital. There are therefore many levels of semi-mechanization for renovations and adaptation in the lime and pozzolana industries.

For a small-scale mechanized lime factory, the winning of the limestone and conveyance to the factory will be carried out through mechanical means. Breaking of the limestone to manageable sizes is carried out mechanically. Normally, vertical shaft kilns are used. These types of kilns are intermittent and have varying sizes. In such kilns, the burnt lime is taken from the bottom at regular intervals and fresh charges of limestone and fuel alternately fed from the top. The lime is slaked in shallow concrete tanks, sieved and bagged.

For a semi-mechanized pozzolana factory, the naturally occurring pozzolana is won with excavators and dried in a drum kiln before grinding in a ball mill. The ground pozzolana may be bagged or mixed with cement or lime to form pozzolana-cement or lime-pozzolana.

When the source of the pozzolana is artificial or from agricultural waste, the calcination is carried out in specially designed incinerators. The ash is further ground to cement-fineness and bagged for use.

Capital-intensive technologies

A fully mechanized technology is capital-intensive and very sophisticated. As much as possible, all processes that are undertaken manually in the semi-mechanized and labour-intensive technologies are mechanized.

For the lime and pozzolana industry, rotary kilns will be required. A mechanized hydrator will be required for slaking the lime and a ball mill for grinding the pozzolana. Homogenizers and silos will be required for mixing and storage before bagging. The capital-intensive lime and cement industry, including pozzolana, uses computers for controlling production. The capital investment for such a set-up is generally huge and beyond the means of many African countries. Presently, this level of technology may not be suitable for Africa. In many African countries, fully mechanized building materials factories have not proved very successful for many reasons including lack of sustained demand, heavy investment and high transportation costs.

Modalities for technology transfer

Information about the range of available technological alternatives, their precise characteristics, and their implications for the criteria of appropriateness is essential for policy making. The criteria for selecting the most appropriate technology have to be identified before any technology transfer is made. This section will discuss briefly the modalities for technology transfer. Many sources of technological information exist and are used. These include some relatively inexpensive sources such as technical literature, formal or informal education and training of individuals, as well as licensing of patented production processes and sale of expertise.

The mechanisms for transferring technology may be identified as.

- licensing of patented processes and sales of expertise and equipment
- joint ventures
- management agreements, and
- technical assistance.

Highly secret know-how, registered trade marks, patents and proprietory technology may be transferred to a recipient for a fee. Sometimes technology transfer is effected through joint ventures between the party possessing the know-how and the recipient.

A third method is effected through management agreement. The party requiring the know-how may request technical management of the technology. The agreement may cover training of skilled labour and technical management personnel.

Technical assistance can also be a means for transferring technology from one party to the other. Normally, technical assistance is arranged government-to-government or between government and any international bodies, such as United Nations agencies which will send experts to the government to advise on specific areas where local expertise is lacking.

The outright purchasing of licences and patents from one party by the other presupposes that the recipient has adequate personnel with technical know-how who will understand the technology and apply it in producing the required products. In the joint venture mode of transfer as well as the management agreement mode, personnel who will operate and manage the technology may be lacking by the recipient organization, so the party transferring the technology will have to train the recipient.

Training rural or urban communities to adopt a certain way of doing things is another mechanism for transferring technology. That kind of technology transfer occurs between a national institution such as a building research organization and the communities of the country and it is usually tied up with a project.

To benefit from such transfer of technology, the recipient must monitor and regularly review and evaluate the performance of the technology so that adjustments may be made where necessary.

Uses of lime and pozzolana

There are many uses for lime, such as in agriculture for improving soil condition. It is used in the paper and sugar industries and also for construction, as mortar or plaster or for soil stabilization in road construction.

The main uses of pozzolanas are for lime-pozzolana mortars, blended pozzolanic cements and as an admixture of concrete mix. Lime-pozzolana mortars are water-resistant and have been, for a long time, the only known cements suitable for such exposure. There are many Roman monuments and ancient monuments in Egypt and India standing today which were built with lime-pozzolana mortar.

Pozzolana Portland cements have been found useful for improved durability, combined with some cost savings, in concrete for marine, hydraulic and underground structures. Their greatest use is in mass concrete work where heat of hydration is reduced with consequent elimination of excessive cracks.

Pozzolana cements inhibit expansion due to the aggregate-akali reaction. They are also used to increase the resistance to sulphate attack or to leaching by soft water. Pozzolana cement is suitable for most non-reinforcement work. Using these two cementitious binders in construction will extend the supply of cement, thereby releasing more cement for other construction purposes.[4,5]

Constraints affecting lime and pozzolana

The earlier sections of this paper tend to paint a rosy picture of favourable factors for promoting indigenous building materials production, especially lime and pozzolana cements. The fact that, currently, the lime and pozzolana cement industry has not been well established on the African continent means that, despite such favourable factors, there are also constraints limiting their production and use. These constraining factors may be broadly classified as:

- those that relate to the importation of building materials, and
- those that relate to the production of building materials locally.

Of these broad classifications, some of the constraints have been identified as basic to most developing countries, including Africa, and are responsible for the inability of indigenous materials to make an impact on the building materials industries. These constraints are related to at least four issues, namely:

- technology of production
- investment requirements
- demand for indigenous products
- inappropriate use of materials in construction.

Importation of building materials

The construction industry in Africa is dependent largely on imported building materials including ordinary Portland cement. The slow pace of growth of local production has not helped to reduce this dependency which has become increasingly difficult for many countries to finance. In some countries well over 90 per cent of the value of building materials used in the formal construction sector is accounted for by imports.

In the face of severe economic problems and the increasing cost of imports, due partly to energy crisis and rising worldwide inflation during the past decade, many countries have had to reduce drastically their financial allocations for building materials imports. Many African countries at the moment cannot afford to devote their meagre foreign earnings to the import of building materials to meet local needs.

This historical dependence on imported building materials has been a constraint to increasing availability and use of locally produced building materials. For example, by depending solely on imported clinker, the development of the lime and lime-pozzolana and composite-cement industry has been ignored.

Local production of building materials

The second major category of factors that have restricted the growth of building materials supply relate to the local production of such materials. Many African countries have established plants to produce cement, lime, burnt bricks and tiles, as well as steel, to reduce dependence on imports and to increase local self-sufficiency in the supply of such materials. The performance of many of such industries has, however, fallen short of expectation, and installed capacities have never been fully utilized. For example, a survey conducted by UNECA in 1982 on cement plants in Benin, Burkina Faso, Ghana, Cote d'Ivoire, Liberia, Nigeria, Senegal and Togo showed that for integrated cement plants the rate of capacity utilization in 1982 ranged between 15 and 90 per cent with an average of 69 per cent. The capacity utilization in grinding plants for the same year ranged from 33 to 87 per cent with average of 57 per cent.[5,6]

The factors underlying the apparent poor performance of building materials plants in African countries are many and varied but a few may be mentioned:

- Inadequate project conception including use of inappropriate technology resulting in high operation cost with consequent loss in profitability and high selling prices which sometimes, due to over-estimation of demand at the planning stage, has also led to low capacity utilization.
- The problem of lack of foreign exchange to buy spare parts and sometimes raw materials from abroad to keep plants running is linked to the type of technology adopted. A pre-requisite for commercial production of indigenous building materials therefore is that the technologies involved be tested, proved and, above all, widely known at the local level.
- High energy cost is another constraining factor. Building materials industries are by nature energy-intensive. The building materials industries were hit particularly hard by increasing energy prices. In response to this, intensified efforts have been made to reduce energy consumption, improve the efficiency of energy use and switch users from oil-based fuels to locally available and preferably renewable sources.
- Poor management practice and lack of skilled manpower. The effect of this has been particularly evident in cases where turn-key projects based on capital-intensive technologies were handed over to local ownership.[5,6,7]

In addition to the poor performance of plants that have been established, other factors have contributed to the slow growth of the local building materials industry in African countries:

- lack of decentralization and diversification of production
- investment requirements
- lack of adequate standardization and quality control
- outmoded building codes and regulations
- insufficient demand for indigenous materials
- inadequate research and information infrastructure
- inadequate policy, planning and promotional support.[5,6,7,8,9,10]

Another matter relating to government policies for the building materials industry concerns the breaking down of psychological barriers against the use of non-conventional building materials. The problem of acceptability crops up with the use of lime and lime-pozzolana and stabilized soil, etc. Firm policies on the use of such materials backed by a demonstration of the government's commitment to use them in government-sponsored projects would contribute greatly in enhancing their increased application in both public and private housing and building projects. It is important also that African governments ensure that priority projects, such as the development of indigenous building materials that are expected to make sufficient contributions to an improvement of the supply of essential building material, receive priority in the allocation of funds.

Action programme for producing lime and pozzolana in Africa

It is revealing from the foregoing discussions that African countries have the potential, in terms of raw materials availability, to increase local production of lime, pozzolana and pozzolana cements, so as to reduce the high dependency on foreign sources of building materials supplies. The discussions have also revealed constraining factors which militate against local building materials production in African countries. It is important to note that the success of the development of the indigenous building materials programme depends largely on the seriousness with which African governments embrace the programme.

It is important that they renew their political commitment to the formulation of clearly defined policies and long term programmes for the building materials industry, coupled with financial allocation to ensure achievement of targets of the programme.

To achieve increased lime, pozzolana and pozzolana cement production in African countries, the following action plans have to be followed:

i) Identification, investigation and documentation of limestone, clam shells, pozzolana and clay deposits, including small-scale deposits.
ii) Information on both large and small-scale deposits of limestone, pozzolana, clay, etc., should be made available.
iii) Plants of different sizes using locally available materials and suitable technologies, tailored to the demand of defined geographical areas should be judiciously sited to help in providing additional cementitious binders from lime and pozzolana.
iv) Technologies that lend themselves to technological adaptation and renovations are to be preferred.
v) Standards and specifications should be formulated for lime, lime-pozzolana and composite cements, and their use should be enforced. Building codes of practice and regulations regarding locally produced materials should also be formulated.
vi) Development of workforce and training to provide managerial and technical skills to cope with requirements of technologies adopted. Also training of skills in use of indigenous building materials should be undertaken.
vii) Development of an effective building research and information infrastructure to provide suitable technologies, assist in standardization and quality control, formulation of appropriate building codes and regulations and dissemination of information to various user-groups.
viii) Strengthening research co-operation among Building Research Institutes in Africa to facilitate the sharing of experiences in the field of building materials development and utilization.

ix) Governments of African countries should use indigenous building materials for constructing schools, offices, clinics, health posts and bungalows, etc.

SECTION II

Production of lime

This section is intended to provide a guide to the current technology for producing lime and other alternative binding materials. It reflects practical achievements made in the past ten years which might possibly be successfully implemented elsewhere in the world. It was not the intention to accept and include reports on research and development on the laboratory scale or those that have not been found to be feasible in practice. In some cases, though, where laboratory test findings appear to contribute new knowledge, open up additional materials for study or have good potential for further development and successful implementation, these have been included. Such a case is the chapter from Vietnam on the pozzolanic reactivity of basaltic rocks, having silica contents under 50 per cent, which were fired at 500 to 800°C.

There is no information about other binders such as gypsum and sulphur; the focus of the Production section is lime and pozzolana raw materials, lime kilns and lime manufacture. Two other reports deal with a hydration plant and making the refractory bricks for lining small shaft kilns. The main development work on rice husk ash 'cement' had been done by the late 1970s, the results of which are already well documented, but a decade later there are no known examples of successful commercial manufacture. The chapter by R.G. Smith is a timely review of the technology as it is now known and of the current situation in ten countries.

In the 1970s, the main thrust in the development of small oil-fired lime kilns was in Indonesia by institutions based in Bandung. Here again, though, there has been no news confirming that those projects have now resulted in assimilation of the Bandung technology by the lime industry creating economically successful new production units.

On the other hand, the development work done by both the KVIC (P.H. Naik) in India and SIDO (E.G.S. Ikomba) in Tanzania has resulted in continuing lime manufacture with, in the case of KVIC, the dissemination of its designs to many countries. During the past few years it has been ITDG which has been amongst the most active in smaller-scale lime manufacture, with its development of kiln and hydrator technology in support of the lime producers who are based on the 'difficult' local Precambrian limestone and endeavouring to supply the needs of the Malawian sugar industry. Spiropoulos and Jones provide details of the technology that has been developed there.

Of the eight chapters dealing with raw materials, one is on testing limestones in the field (N. Hill), one is on dolomite as raw material in Sir Lanka (J.W. Herath), three deal with volcanic ash pozzolana (W.J. Allen, R. Tuts, W. Balu-Tabaaro), one describes tests on a pozzolana made from basalt (N. Van Minh), two include rice husk ash pozzolana (R.G. Smith, R. Tuts) and one is on sugarcane ash as a pozzolana (S.F. Ahmad & Z. Shaikh). There were five chapters dealing mainly with design and operation of lime kilns (J. Pacheco, P. Nhachi, E.G.S. Ikomba, J. Spiropoulos, P.H. Naik), one on making refractory for lime kilns (C. Stevens), one on a hydration plant (B. Jones) and one on the lime situation in Algeria (J. Moussa).

Neville Hill

6. Field-testing of limestones for limemaking
NEVILLE HILL*

Projects for large-scale manufacture of cement, lime, bricks and other building materials, where the capital investment is likely to be in the tens of millions of dollars, have to provide for detailed testing of the proposed raw materials so as to be sure that they will perform satisfactorily during processing and produce good quality products. It would be quite normal for the project engineers to insist that a bulk sample of several tonnes of the limestone or clay is tried out in a full-scale process, which may have to be, say, on a plant in Europe. This would be in addition to complete chemical testing, laboratory firing trials and other physical tests. Such a testing programme could contribute 5 to 8 per cent of the total project cost but would be fully justified as being an essential insurance against the value of the investment in plant, infrastructure, etc. being lost because the raw material source proved to be unsuitable.

The investment cost, though, of the smaller-scale lime and brick-making projects is normally in the tens of thousands of dollars, rather than tens of millions. There is seldom much hope of getting funding for doing a complete raw material study including full-scale plant trials. Instead there has to be compromise with enough testing being done so as to be reasonably sure of selecting the best of the available raw materials. Instead of mining and transporting a large quantity of material for trials in a laboratory abroad and collecting many samples for analysis and testing at a national geological survey laboratory, it is very useful, instructive and time saving to be able to carry out preliminary small-scale testing in the field.

The professional training of geologists and chemists tends to instil in them the need for careful, detailed investigation in their work. They usually want to insist that a lengthy and possibly costly 'complete' assessment programme is appropriate. But their natural desire to have an 'exact' result has to be counterbalanced by the needs of the local entrepreneur or cooperative who are wanting to get on without too much delay or added expense. All they really need to know is that there will be enough of the raw material (and at an acceptable cost into the plant) and that it will produce a saleable product.

---

*TERRE, Portsmouth, UK.

This chapter describes the building and use of a small kiln for doing test firings of local limestones with a view to establishing lime manufacture in the far north of Malawi. The author was collaborating in 1981-2 with the Malawi Geological Survey Department (GSD), based in Zomba in a UNIDO project aimed at compiling an inventory of ceramic and other raw materials including limestones and pozzolanas. A means of eliminating, whilst the party was still in the field, those deposits of limestone that were clearly unsuitable would reduce the number of samples to be taken back and tested in Zomba. In addition, being able to do testing locally would reveal the deposits at which it would be worth doing more detailed study and sampling.

Firing lumps of limestone in a small kiln and observing the results is a similar approach to that for assessing clays for brickmaking by building a small clamp to test-fire bricks in the field. Whilst in Karonga, the author remembered John Parry's words to a meeting of ITDG's Building Materials Panel – 'If you want to know if a clay is any good for brickmaking, make some bricks and burn them!'.

Building materials in the north of Malawi

Karonga and Chitipa Districts are located at the northernmost extremity of Malawi, bordering the frontier with Tanzania. The area is the furthest from the cement factory at Changalume in Zomba, around 600km to the south, and from the then lime-producing areas at Lirangwe in Blantyre and Chenkumbi Hill in Machinga. The retail prices of cement, lime and other construction materials were very high in the two districts despite a price equalization policy operated by the cement company for sales from its regional wholesale outlets at Blantyre, Lilongwe and Mzuzu. The retail price of Portland cement in Karonga was expensive at the equivalent of US$170 per tonne, but white hydrated lime was double this price at US$335.

There was clearly scope for establishing local production of white lime in the north to replace the lime transported in from some 500km to the south

Table 1. **Building material retail prices (Hardware & General Dealers Ltd) July 1982**

Material	Quantity/ Quality	Zomba	Mzuzu	Karonga
Portland cement	50kg	7.64	7.10	8.98
Hydrated lime	25kg	4.25	4	8.80
PVA white emulsion	5 litres	15.85	–	18.65
Galvanized corrugated iron sheets	28 gauge	–	–	3.11/metre
	26 "	2.82/metre	–	3.79/metre
	24 "	3.14/metre	–	–

(Prices in Malawian Kwacha; MK1.00 was US$0.94)

and so provide a low-cost white decoration as an alternative to imported PVA paint. Lime was also said to be needed for soil stabilization in a new road project between Karonga and Chitipa, and by state-run farms for adjusting the soil pH – though crushed limestone is usually more appropriate for doing this. (In view of its high price compared with cement, no lime could be expected to be used in other building work in the area.)

The coal at Ngana, which has an average calorific value of 22MJ/kg and an average ash content of about 22 per cent, had been shown to burn well during a previous small-scale limestone firing trial conducted in the 'Zomba' kiln which had been built at the GSD's headquarters. (This kiln design was also used by John Spiropoulos in his initial work for setting up lime-making for the Southern Rural Development Association in Botswana.) Possible local sources of pozzolanic material included the Songwe Volcanics – pumiceous and fine-grained tuffs south of the Songwe River with a little occurring as far east as near Karonga Boma – and rice husk ash (RHA) from the Chilumba rice mill on the shore of Lake Malawi.

Locally, rural houses are constructed mainly of rammed earth with a roof of either thatch or galvanized corrugated iron sheets.

The 'Karonga' kiln

Materials

The materials for the small test kiln built and used in northern Malawi were those that fortunately just happened to be lying around in the compound of the district depot of the Lands Valuation & Water Department which the GSD team was using as a site for its base camp during four weeks of geological field work in Karonga District.

The useful items found were 256 fired clay bricks, a clayey soil to make a sticky mud 'mortar', some discarded lengths of 19mm diameter and 41mm drill rod and rectangular section iron bars 63mm × 9mm which could be sawn up to make the grate etc. A 'luxury' item, which could not normally be expected to be present except in association with water drilling equipment, was the casing for a Coventry Climax water pump. This proved to be just right for the chimney.

Construction

The kiln had double thickness brick walls making up three sides of a 0.75m square and was stood directly on a patch of firm level ground. A shallow pit was opened nearby for extracting the clayey soil to make a stiff mud mortar. There were 16 bricks on each course and the kiln was 14 courses high or about 1.20m and, with the chimney, had an overall height of 1.50m. After laying the first two courses, lengths of 20mm drill rod cut to 0.75m were laid across the width of the kiln. These would support the burning wood and provide an ash pit underneath. After a total of six courses of bricks, three lengths of 40mm drill rod were laid horizontally and fairly close together across the front of the kiln to provide a support for the bricks which would eventually close the kiln after it had been loaded. To the rear of these rods were four lengths of the rectangular iron bars stood on edge to provide the support for the limestone.

Construction continued with the bricks placed directly above each other for the first 11 courses. Except at the front, the 12th and 13th courses were slightly corbelled inwards nearly closing in the top of the kiln, which was done by spanning the gap with the discarded water pump casing. An extra layer of mud and bricks was placed around the chimney to seal gaps and reduce escape of heat during firing. The capacity of the kiln, or volume in which limestone could be placed, was about one cubic foot ($0.028m^3$). Nearby a shallow slaking pit was made, one brick deep and lined with clean bricks.

Operation

The limestones being tested were broken into lumps of about 70mm size and filled the kiln. All the wood fuel was burned in the grate beneath the limestone. Firing was usually started by 0700hr and continued to around 1700hr when the kiln was allowed to cool down overnight. A note was made of any phenomena such as explosive sounds and the physical state of the fired material such as proportion of fine fragments produced. During these tests no measurements were made of the weight of limestone, before and after firing, but provided the rock did not decrepitate much it would be possible, for instance, to determine the weight loss and hence the percentage of original rock converted into quicklime.

The quicklime was transferred to the small slaking pit and sprinkled with clean water. A note was made of the time until some reaction was seen; the vigour of the reaction (slaking) and the nature and proportion of any unslaked material was also recorded. The lime putty was made into a creamy 'paint' with more water and brushed onto a convenient surface and allowed to dry. At Karonga spare house bricks were laid out as an exposure site, but any convenient surface such as a smooth wall of rammed earth could be used.

The simple step of checking on the 'whiteness' of the resulting limewash immediately indicates which limestone deposits would be unacceptable for industrial uses, such as in the sugar industry, or as decoration. The colour of the original limestone is often no indication of how the hydrated lime will appear. Thus the pale to medium grey Mwesia limestone produced a muddy, yellow-brown lime putty. The famous limestone of Carboniferous age used by ICI for limemaking in Derbyshire, England is a medium or darker grey and yet the hydrate is white and ideal for limewash as well as industrial uses.

A further simple test would be to check on possible unsoundness, which could result if the limestone were dolomitic, by observing if any 'pitting and popping' occurs on the surface after the lime putty is applied as a plaster; this would be caused by the late hydration of any magnesia, MgO, in the plaster.

Testing some limestones of the Karonga area

Whereas the Malawi lime industry in 1982 was using Precambrian 'Basement Complex' marbles, i.e. metamorphic rocks concentrated in two or three areas in the south, the principal occurrences of sedimentary limestones are all in the north, in Karonga District. During the field work, four limestones were inspected and sampled and were subjected to test firing in the kiln and subsequent slaking.

Mwesia, near Mpata

The limestone with apparently the best field characteristics as a raw material for limemaking was that in the Mwesia area, near Mpata. This is a Karroo sediment (Permo-Triassic), medium to pale grey, hard, fine grained, slightly oolitic limestone which occurs as blocky beds up to 0.6m thick. There would be plenty of material to feed a continuous kiln consuming, say, 10 tonnes a day.

But in fact, the firing and slaking trials were disappointing. The rock shattered explosively during the first 2 hours of firing and many fragments fell through the grate. Upon opening the kiln, after 10 hours adding fuel and 24 hours cooling, there was no powder but many small fragments besides the few unbroken lumps and the draught through the bed would be much reduced.

The quicklime was 'quick slaking', several pieces steaming and breaking up in less than 5 minutes. The colour of the 'whitewash' upon drying was buff or pale yellow-brown. After 20 minutes hydration, there still remained some unreacted medium grey, fine grained stones, possibly of a silicified limestone.

Ngana area

The limestone here is also is a Karroo sediment and is medium grey, hard, fine grained, compact, sometimes siliceous. It occurs as strong beds about 0.5m thick with some softer, yellowish marly layers.

Upon firing this limestone also exploded, though not as much as the Mwesia rock, and on opening the kiln some decrepitation was seen. The lime hydrated fairly actively, with some popping and steaming within 5 minutes. Several lumps were left that did not hydrate and which were hard and siliceous. The colour of the lime putty was creamy yellow and though not 'white' when dry it was light enough to possibly attract some use as a coating for walls of rural houses.

As most Karroo limestones in Karonga District are argillaceous, limes made from them should be hydraulic to some extent, so the lime putty was mixed 1:3 with clean quartz beach sand from the lakeshore near Karonga. This mixture was applied as a 10 to 20mm topping over an area of about 1 metre square of compacted reject limestone chippings and left covered with a plastic sheet. (There is no record of how this performed.)

Uliwa area

Samples were collected from two lithologies of the Chiwondo Beds limestone (Miocene? age) which both outcrop within a mile of Uliwa traditional court. The lower limestone horizon is conglomeratic to sandy limestone with small pebbles and coarse sand-sized grains of quartz and chert set in a ground mass of fine-grained calcium carbonate. This could be a calcrete, i.e. a gravelly calc tufa caused by a spring of water, charged with dissolved lime, passing through a bed of sand or fine gravel.

A creamy-coloured, medium hard, fine grained, platy, somewhat marly limestone occurs higher up. The bed thickness is about 5m and the reserves estimated to be about 20 000 tonnes or enough for establishing a 10t/d kiln.

Both types performed well in the kiln; when it was opened the quicklime was still in its original lumps with little or no decrepitation. They slaked fairly rapidly to a fine, white powder. The whitewash appeared 'white' and compared very favourably with the best lime made from the Precambrian marble in the south at Chenkumbi.

Conclusions

The small-scale trials with the 'Karonga' kiln provided useful findings on the potential for limemaking of the limestones in the north of Malawi.

Building such a kiln can be done in less than a day and samples from prospective deposits can be tested at the rate of one a day. Apparently good looking limestone resources in sedimentary Karroo limestones at Mwesia and Ngana, for instance, were eliminated by the test firing and slaking whilst on the other hand the younger Miocene(?) limestones in the Chiwondo Beds, such as occur around Uliwa, performed well and have been used for limemaking in a subsequent project.

Geologists and others carrying out a search for suitable raw material for a lime project should keep a look out for any materials that would be useful for making a small kiln. Scrap yards and building sites are worth visiting – with permission, of course!

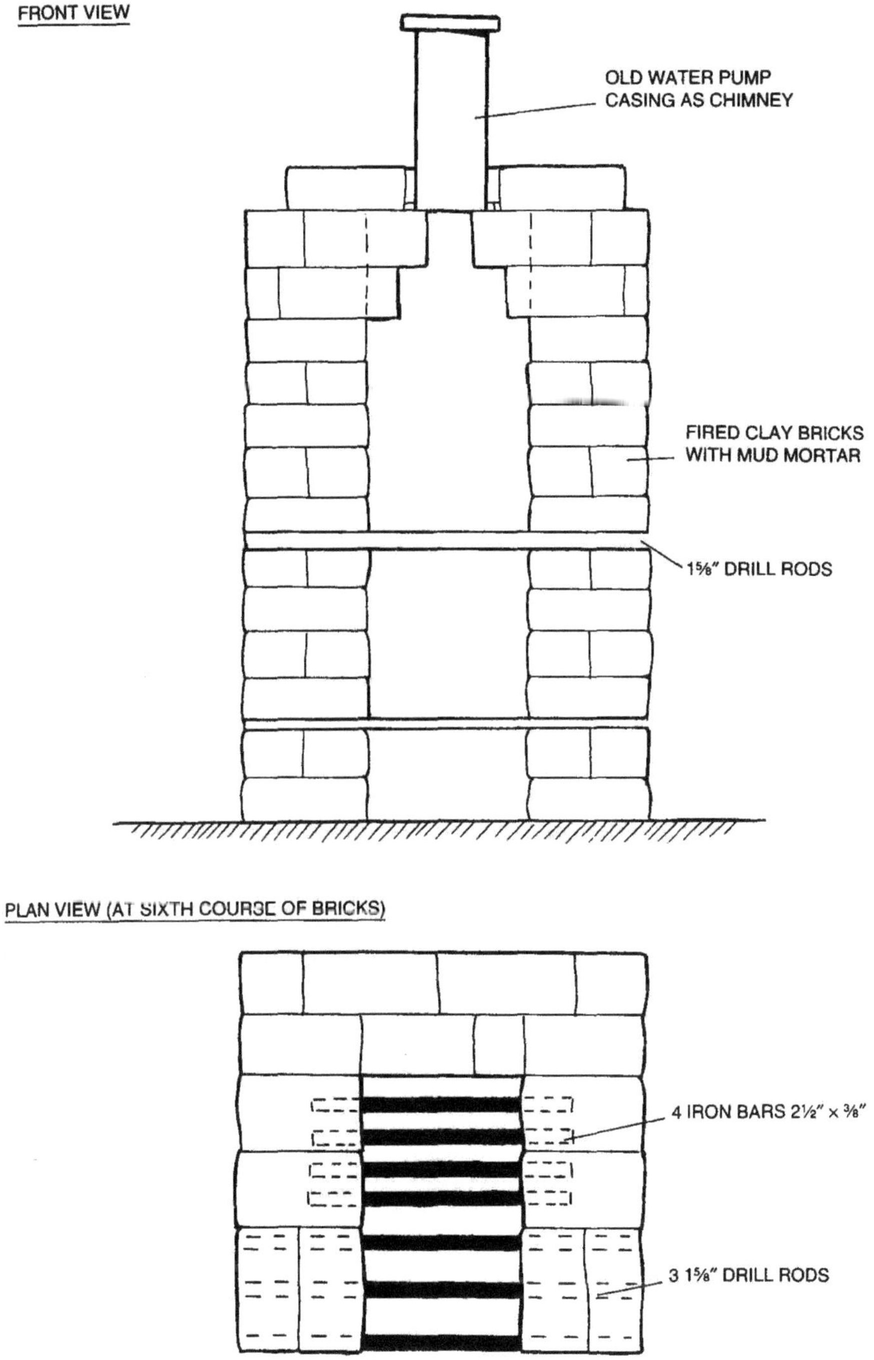

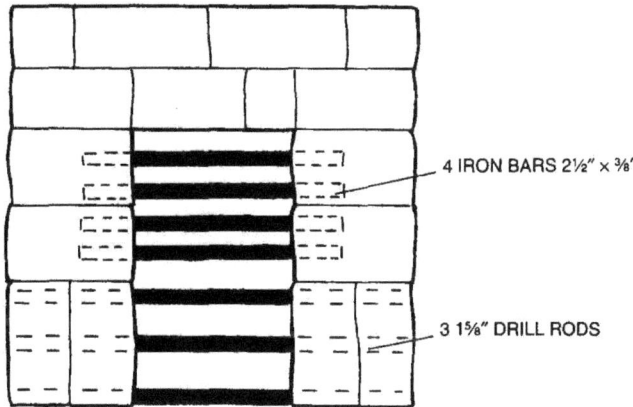

Figure 1: *Karonga kiln – a small batch kiln used for test-firing of limestone samples*

7. Lime industry in Sri Lanka
J W HERATH*

In Sri Lanka, the cottage lime industry, which supplies over 80 per cent of lime to the building industry, is mainly confined to coastal areas, utilizing coral and shell formations as the raw material. In the western coastal belt, the Akurala–Hikkaduwa area is noted as a centre of the cottage lime industry.

In recent years, because of destruction of coral reefs, the state has banned mining off-shore and up to 300 metres inland from the coast. This restriction of coral mining has affected the cottage lime industry requirements of raw materials for lime production although inland mining is permissible beyond 300 metres from the coast. The question has now been raised as to what alternative sources of calcareous material are available for lime production both in already restricted areas and in other areas of the island.

Outline of geology and mineral resources

Over 90 per cent of the country, covering an area of 65 600km with a population around 16 million, is underlain by Precambrian rock (high grade metamorphic rocks). Associated with these rocks are granites and granitoid rocks of igneous origin. Jurassic rocks are limited in extent. Miocene limestone extends from the Jaffna Peninsula in the north along the north-west coast. Included in the Quarternary Period are the Pleistocene deposits and recent formations.

Sri Lanka cannot be classed as a major producer of mineral commodities. The country is however endowed with industrial minerals and rocks and these are being exploited and utilized to a considerable extent in an already established mineral-based industry. All minerals exploited are of the non-metallic type. An account has been given of the mineral resources of Sri Lanka.[1,2,3,4]

Raw materials for lime production

The term limestone is applied to many forms of calcium carbonate, each with distinct physical properties. The differences between them are marked and yet, when they are separated from any small percentage of impurity present and are then carefully studied, it will be found that they all consist of calcium carbonate. In Sri Lanka limestone occurs in various forms and these varieties include:

---

*Ceramic Research and Development Centre, Piliyandala, Sri Lanka.

- The Miocene limestone (over 95 per cent $CaCO_3$)
- Calcite (100 per cent $CaCO_3$ – pure)
- Coral (over 95 per cent $CaCO_3$)
- Shell (over 95 per cent $CaCO_3$)
- Dolomite (MgO variable)

Table 1 shows the chemical analysis of various calcareous materials (Herath 1985). Sedimentary limestone deposits of the Miocene age are best developed in the Jaffna Peninsula where they occur as a hard compact limestone. The over-burden at some points may be over 30 metres thick. Chief impurities in the limestone are varying amounts of clay and silica and traces of magnesia. Two cement factories have been established in the area, one at KKS and the other at Puttalam.

Table 1. Chemical analysis of limestone

Constituents	Shell	Coral	Miocene limestone	Dolomite	Calcite
SiO_2	1.15	2.00	0.82	0.75	Tr.
Al_2O_3	0.41	3.40	0.52	0.27	Tr.
Fe_2O_3	0.33	0.50	0.08	0.05	–
CaO	54.89	51.50	54.20	31.01	55.23
MgO	0.02	–	0.70	21.78	–
Loss on ignition	43.15	43.68	43.68	46.10	43.83
TOTAL	99.95	101.08	100.00	99.96	99.06

The proven reserve of Miocene limestone is in the region of 100 million tonnes, sufficient for a period of nearly 30 years at a production rate of 2 million tonnes of cement per annum. The inferred reserves in the area are very much larger. Miocene limestone is not very popular for lime production due to its off-white colour.

Coral deposits are found at various points along the coast of the island. They are not of great importance except as a raw material for lime manufacture by the cottage industry along the stretch of coast from Ambalangoda to Matara. The deposits consist of loosely-packed finger or stock coral with heavy blocks of massive coral. Coral deposits are also found overlying the Miocene limestone at the margin of the Jaffna Peninsula in the north. Other areas noted for coral beds include Kalkudah, Kachchaveli and Delft Island.

Coral mining off-shore has caused serious erosion at several points along the coast. This activity has now been banned by the State. It is estimated that around 25 000 tonnes of lime per annum are produced in the region from coral as raw material.

Shell mining is mainly concentrated in the Hambantota District. Extensive deposits of sea-shells occur at Hatagala (Hungama). These deposits stretch parallel to the coast and are about 3km wide.

The shells occur to a depth of 2 to 3 metres from the surface and are mixed with clay, silt and other impurities. Although considerable amounts have been mined, the reserves in this area are still estimated at a million tonnes. Large quantities are transported to Colombo and other areas. A modern lime plant has been established by Lanka Ceramic Limited at Hungama. Mining of shell is also prohibited in the Wild Life Department reserves. Shell mining is limited when compared to coral mining.

Calcite deposits occur in some areas in the island and the best known are in the Balangoda area. Calcite deposits up to now are used mainly in the ceramic industry and the resources available are small and cannot be considered for lime burning. These limited deposits are of high purity.

The inferred reserves of dolomite in Sri Lanka are over 500 million tonnes (8 to 21 per cent MgO) and are confined to the belt of marble (dolomites) in the central hill country. Dolomite with an MgO content of 18–22 per cent is of common occurrence. Accurate records of the annual production of dolomite are not available but it is estimated to be around 25 000 tonnes of crushed material per annum and 15 000 tonnes for lime production. All dolomite deposits are found at or near the surface in readily accessible areas so that simple quarrying or open-cast operations are all that is required. The capital required is small and mining of dolomite has potential for employment on a large scale. Dolomite with its numerous applications has also great export potential. Dolomite is used in Sir Lanka for the manufacture of ceramics, glass, fertilizer and in a number of other industries. Dolomite can also be used for refractories and they make handsome building stones. It is also used as a source of lime in the absence of

Table 2. Limestone reserves and mineral supply position for industry ('000 tonnes)

Commodity	Reserve (measured)	Reserve (indicated)	Reserve (inferred)	Mineral supply position from measured and indicated reserve
1. Miocene limestone	100 000	50 000	200 000	30 years
2. Coral (inland unrestricted area)	Not available	300	400–500	6 years
3. Shell (unrestricted area)	Not available	100	150	8 years
4. Calcite	Not available	30	60	20 years
5. Dolomite	Not available	200 000	500 000	Unlimited over next century

limestone deposits, mainly in the hill country of Sri Lanka. Table 2 shows the limestone reserves of the country and the mineral supply position for industry.

Quicklime or calcium oxide (CaO) is produced by heating limestone or calcium carbonate ($CaCO_3$) and most quicklime contains 95 per cent or more of calcium oxide. When quicklime is exposed to moist air, or treated with water, it slakes, forming calcium hydroxide $Ca(OH)_2$, slaked lime or hydrated lime.

100 parts by weight of calcium carbonate in the process of decomposition will result in 56 parts by weight of lime and 44 parts by weight of carbon dioxide. Overheating must be avoided as it forms lime which does not slake satisfactorily. Highly calcareous limes are chiefly used for chemical purposes and where a relatively pure calcium oxide is required.

A magnesian lime is one containing a noteworthy proportion of magnesia. Magnesian limes are produced in the same manner as other limes. When a dolomite or dolomitic limestone is used a dolomitic lime is produced as in the hill country of Sri Lanka. The magnesium carbonate is converted into the oxide (MgO) at a slightly lower temperature than calcium carbonate so that the magnesia in a magnesium lime could be overburnt with the result that it slakes very slowly.

In a dolomite ($CaCO_3.MgCO_3$), the $MgCO_3$ dissociates to MgO at 775°C. The dolomite when heated to 875°C will result in the overburning of MgO. When dolomitic lime is used for building purposes the overburnt MgO is not readily converted to the hydroxide $Mg(OH)_2$. If this material is used without allowing the hydration to be completed, blistering of plasters occur when dolomitic lime is used for plastering walls. It is therefore desirable to leave dolomitic lime after hydration for long periods for the completion of MgO to the hydroxide. If it is used for mortar it need only be kept for three days after slaking.

The behaviour of calcium carbonate, when heated, depends on the pressure of the carbon dioxide present. Magnesian limestones ($CaCO_3 . MgCO_3$) behave differently. In certain instances, the whole of the magnesium carbonate is first decomposed and the calcium carbonate later; but other magnesian limestones, when burnt, form lime (CaO) and magnesia (MgO) simultaneously. The latter are normally the true dolomites. Excessive pressure of carbon dioxide gas in a kiln may sometimes result in recarbonation ($CaO + CO_3 \rightarrow CaCO_3$). The decomposition of limestone can therefore occur when the pressure of the carbon dioxide gas is sufficiently low. Complete and rapid decomposition will occur if the gas is removed rapidly as it forms so that pressure in the kiln never approaches that of the atmosphere. The removal of the gas is most easily effected by drawing a current of air through the kiln. An excessive draught will do no harm if the kiln is sufficiently tall. In short kilns, excessive draught will carry away much heat and will be wasteful in fuel. An alternative method of preventing the pressure of the CO_2 becoming excessive is to inject steam into the kiln. The use of air is, however, much cheaper. Some lime burners wet the stone before putting it into the kiln. This procedure is also somewhat wasteful in fuel.

In the hill country area of Sri Lanka where magnesium limestone or dolomite is used for lime production, continuous vertical kilns are in operation. Masons and plasterers are experts at working with dolomitic lime. Dolomitic lime to be used successfully should be added to water (not water to lime). The material is then stirred in a shallow well and the coarse material settles at the bottom. After some time the surface fine material is recovered for plastering purposes and the material at the bottom is used for mortar. A large number of houses in the hill country do not have any cement in their construction. If cement is used it is in negligible amounts (15 parts to 1 part of cement in mortar). Builders claim that where lime mortar is used without cement the mixture gets rock hard after three days. Cement, if used at all, is only used for concrete mixtures. There is, therefore, a great advantage in using magnesium limes as a considerable amount of cement could be saved in the process of building. The problems for processing dolomite for lime in Sri Lanka are the lack of dynamite for mining and the scarcity of fuelwood for burning. All lime kiln operators are very much concerned over these two areas but they are helpless.

Table 3 is presented as a comparison of properties of dolomitic and calcium limes. Although it is believed that long periods of slaking are necessary for dolomitic limes before use, masons and plasterers are of the view that this is not so with dolomites from the region. What has to be remembered is to add dolomitic lime to water when slaking and not water to dolomitic lime. With Sri Lankan dolomites used for lime manufacture it may be that the $CaCO_3$ and $MgCO_3$ in the dolomite dissociates around the same temperature or within a very narrow range of temperature (say 15–20°C). It is considered that this may be the reason that blistering of plasters does not occur when dolomitic lime is used. Considerable years of experience are therefore required to operate a lime kiln for quality lime production.

Lime production in Sri Lanka

Lime production in Sri Lanka dates back to antiquity and is also one of the oldest technologies in the island. The industry is mainly concentrated in three regions:

Region 1 – North-western, Northern and North-eastern parts of the island.
Region 2 – South-western and Southern coastal areas.
Region 3 – The hill country.

Table 3. Comparison of properties of dolomitic and calcium limes

Properties	Calcium limes coral – shell – others	Magnesium limes Magnesium limestones and dolomites
Magnesia content	Negligible Less than 1 to 2% MgO	MgO content varies – common range 15–21% MgO (in Sri Lanka)
Impurities	May contain sand unburnt $CaCO_3$	Unburnt $CaCO_3.MgCO_3$
Slaking	Excellent	True dolomites (MgO 21%) slake satisfactorily like calcium limes. In some case MgO is overburnt.
Plasticity	Excellent	Less plastic than calcium limes.
Smoothness	Excellent	Less smooth (termed weak, lean or poor)
Burning characteristics	Good – as no $MgCO_3$	$MgCO_3$ sometimes is converted to MgO at a lower temperature than $CaCO_3$. MgO in dolomitic lime is sometimes overburnt.
Expansion on slaking	Good (normal)	Less than calcium limes
Shrinkage	Good (normal)	Less than calcium limes
Use for mortar	Suitable	Suitable – gives mortar greater strength than calcium limes
Use for whiting	Suitable	Suitable
Setting time	Good	Slightly slower than calcium lime
Blistering of plaster	No blistering of plaster occurs. Can be used immediately after slaking	Plasterers are afraid of using slow slaking dolomitic lime as blistering may occur on walls or ceiling. Should be left for as long as possible after slaking for blistering not to occur

Region 1

The first two areas are covered by Miocene limestone deposits which are not used extensively for lime production. The cottage industry in these areas prefers to use coral as far as possible. The demand for lime is also not so great as in the south-western parts of the island. The entire region produces around 8–10 000 tonnes of lime per annum. There is no immediate problem of exhaustion of deposits. The north-eastern parts use coral for lime production and some areas use shell with cowdung as a fuel.

Region 2

This region has the largest production of lime in Sri Lanka operated by the cottage industry and one modern lime plant operated by Lanka Ceramic

Limited. Herath (1983) estimated a figure of 22 000 tonnes of lime production in the region from coral and around 8000 tonnes of lime per annum from sea-shells. This would mean an annual figure of 30 000 tonnes of lime from coral and shell. A survey undertaken by the Geological Survey Department in 1985, the most detailed survey carried out so far, identified 303 lime kilns from Akurala to Hambantota. The survey also gave a figure of 48 000 tonnes of coral mined per annum, which produced 26 000 tonnes of lime. The total amount of lime produced from coral and shell is therefore around 30 000–35 000 tonnes per annum.

Region 3
The hill country has to depend on dolomites with varying MgO content (8–22 per cent MgO) for lime production. Reserves are unlimited and quarrying is all that is required to recover the material. No accurate figures are available for annual production levels. A reasonable estimate is around 10 000 tonnes of lime per annum.

The total lime production in all three regions is around 50 000 tonnes of lime, requiring a raw material consumption of about double this amount – 100 000 tonnes. The exhaustion of coral and shell deposits will be felt in another eight years time. From now on it would be desirable to switch to other sources of limestone, e.g. dolomite, for lime production. The State should therefore take the initiative to encourage the production of lime in the hill country for use in other parts of the country. The distance is the same as for coral and shell lime to be transported to Colombo and the suburbs. The other alternative is to concentrate on the use of Miocene limestone for lime manufacture from the Puttalam and Jaffna areas. Miocene limestone is, however, not popular due to its off-white colour.

The prices of lime vary within very narrow limits. In the hill country and in the northern areas lime is sold at Rs1500 to 1750 a tonne (US$1 = Rs42). In the south it is around Rs2000 a tonne and in Colombo around Rs2500 to 3000 a tonne. Lanka Ceramic Limited, operating the largest modern lime factory, markets lime around Rs5000 a tonne. The wage structure for persons employed in the lime industry is shown in Table 4.

The CRDC prototype lime kiln

The CRDC carried out an extensive study on the lime industry of Sri Lanka mainly for the purpose of determining how best the efficiency of lime kilns could be improved.

The following observations were made:

1. The temperature at the top of the kiln at the point where raw material is introduced is always below 100°C. This causes steam to condense at this

point causing a further drop in temperature. The velocity of gases evolved is also affected. This causes the carbon dioxide concentration inside the kiln to increase resulting in a reverse reaction to form calcium carbonate, causing the efficiency of the kiln to drop.

2. The temperature at the centre of the kiln is always higher than at the periphery. In the traditional kiln, the gaseous convectional currents are more concentrated at the centre of the kiln due to its higher temperature. The rate of conversion to lime is therefore faster at the central parts of the kiln. The end-result is that when the burnt product is removed from the bottom of the continuous kiln (layer-wise), lime is found to contain around 20 per cent unreacted dolomite which comes from the sides of the kiln.

3. In the traditional kilns studied, the velocity of the forward reaction cannot be controlled. This reaction depends on the velocity of the wind. It is not possible, therefore, to control the output of the lime kiln.

4. The heat losses in traditional kilns are considerable. The majority of kilns have cracked walls due to poor construction.

The main purpose of the modified kiln constructed by the CRDC was to rectify all the defects observed in the traditional kiln. The prototype kiln was tested and its performance was a great success. The main features of the kiln are as follows:

1. The kiln is 3.6m tall and 1.2m in diameter. The kiln is provided with three walls. Heat insulating materials are incorporated into these, reducing heat losses.

2. There are four chimneys symmetrically placed at the peripheral walls. These chimneys originate from the area of the kiln where the temperature is around 600°C. The CO_2 produced as a result of the decomposition of dolomite is rapidly removed by the chimneys from the sphere of reaction.

3. The chimneys maintain a strong convectional current at the periphery of

Table 4. Wage structure in mining areas

	1983[1]	1985[2]	1990[3]
Labourer (skilled)	35–50	40–55	60–80
Labourer (unskilled) Male	25–45	30–45	50–60
Female	20–30	20–35	35–45
Children	10–20	20–30	30–40
Pump operators (absent in shell and dolomite mining areas)	45–55	60–80	80–100
Kiln operators	45–55	60–80	80–100

Future of the Lime Industry, TIS 5. CRDC 1991, 1–43.
1. Herath 1983 NARA
2. G.S.D. Report No. MR/C/16 – 1985
3. Herath 1990 Field Studies

the kiln enabling decomposition to take place not only in the central parts of the kiln but also towards the inside walls.

4. The upward movement of gases, as well as the forward reaction, could be controlled by dampers in the chimneys.

Summary and conclusions

The lime industry of Sri Lanka is one of the oldest technologies of the country. Both dolomitic and calcium lime were used during ancient times. This industry has been mainly operated as a cottage industry. Sri Lanka has to depend on Miocene limestone, coral, shell and dolomite for lime production. The lime industry has used coral, shell and dolomite as raw material. It is observed that shell and coral deposits will be exhausted in another eight years (no accurate figures are however available for reserves). The lime industry will eventually have to depend on Miocene limestone and dolomite for raw material. Miocene limestone is not favoured as a raw material. The reserves of dolomite are adequate for large scale commercialization of the lime industry. Also dolomitic limes are not much different from calcium limes. The prototype lime kiln designed and constructed by the Ceramic Research and Development Centre has been a great success. The fuel-efficient, continuous and rapid firing kiln is undoubtedly a major contribution to skilful lime production in the country.

8. Locating reactive natural pozzolana
W J ALLEN*

Pozzolana is a valuable material because of its potential to increase the quality and quantity of cementitious binders available to the building industry. It can do this in several ways. It can be used to extend and improve the quality of Portland-type cements. Significant substitutions can be made without loss of strength and with improvements to long-term durability and resistance to chemical attack. This is common practice in the UK and other European economies; a good example being the controlled (by British and European Standards) use of pulverized fuel ash (PFA), a by-product of coal-burning power stations.

It can also be used with lime to confer truly cementitious properties on the mixture. This is the cement used by the Romans in antiquity and promoted in recent years by ITDG among others as an appropriate and

---

*Ellis and Moore Consulting Engineers, UK.

low-cost alternative to Portland cement in applications where the high-strength performance of the latter is unnecessary or undesirable.

Pozzolanas take many forms. They may be industrial by-products such as PFA or rice husk ash (RHA). They may also occur naturally; usually as a consequence of volcanic action. Finally, they may be manufactured – for instance, by the controlled burning of suitable clays.

What are natural pozzolanas?

A pozzolana is a siliceous and/or aluminous material that will, in the presence of water, combine with lime to form cementitious compounds. Naturally occurring pozzolanas may take many forms but most fall within, or are derived from, the generic classifications of glass or clay.

The glasses are usually volcanic products subjected to rapid quenching from high temperatures during deposition. They are amorphous with no defined structure and are usually chemically reactive. Natural weathering processes can convert volcanic glass to crystalline forms which are generally less reactive.

Clays, on the other hand, are crystalline minerals in particulate form, bound tightly together by mutual attraction. They are usually associated with variable quantities of water molecules. Generally less reactive than volcanic glasses, their reactivity is a function of the clay mineral content and type and usually may be improved by heating to critical elevated temperatures.

Reactive versus inactive

Regardless of the intended use of the pozzolana, the most reactive materials will be sought. In the case of volcanic glasses, reactivity derives from the high energy state of the liquid structure of the material frozen in by rapid cooling. It follows that the more rapid the cooling, the more reactive the pozzolana is likely to be. Reactivity is therefore linked to the way in which the volcanic material is formed. Reactivity will be diminished by conversion of the amorphous structure to a more stable crystalline state. This tends to occur as a natural weathering process with time and exposure to the elements. Reactivity can therefore be linked to age and environment.

In most situations, the reactive glass will be combined with some crystalline material and their relative proportion will influence reactivity. The physical particle size or fineness will also play a part by determining the surface area available for the initial chemical reaction. As we shall see, these properties too can be linked to environment and the way the materials were formed.

In the case of clays, reactivity may be related to the type of mineral and the proportion of clay – the clay fraction – in the material. In general, the reactivity of clay as it occurs naturally is relatively low when compared to the volcanic glasses.

Performance may be improved, however, by heat treatment. The effect is to disrupt the well-ordered crystal structure. Temperature and duration of heating are critical as prolonged exposure to too high a temperature can result in re-crystallization and a fall in reactivity. Optimum values depend on the form of mineral present.

Preliminary research

For a professional geologist, it may well be possible to tour an area of interest and, with the benefit of the naked eye, identify potentially pozzolanic materials belonging to the above two categories. Even a professional, however, would benefit from initial research and for the amateur or lay person it is essential.

The most useful starting point is the geological map. This will relate particular classifications of soils and rocks to geographical locations. Unfortunately, like all good documents written by and for professionals, the map and commentary require some translation. The common terms associated with volcanic materials and sedimentary clays are given below:

Volcanic materials

Pyroclastic	Term for materials thrown into the air by a volcanic eruption
Ash	Small fragments from 0.06 to 4mm. May be a mixture of amorphous glassy material and crystalline minerals
Tuff	Consolidated ash
Pumice	Glassy honeycomb rock of low density usually particulate in form
Ignimbrite	A welded tuff

Clays

Argillaceous	Sedimentary rock with a grain size less than 0.0625mm
Clay	Sediment with particles less than 0.0039mm in size. Plastic when wet
Kaolinite	Clay mineral
Illite	Clay mineral
Montmorillonite	Clay mineral

General

Quaternary	Recent deposits less than 2 million years old

These terms may in general be associated with potentially reactive materials. The list is not exhaustive and a good simple geological dictionary will be essential for those with no geological training. The best advice is to consult a professional at this stage but of course this is not always possible.

At this point in the search for a reactive pozzolana, the geological map may be augmented by a good topographical map. Again, those unfamiliar with the conventions employed should consult an elementary textbook on physical geography.

A good map will enable you with some practice, to 'see' the landscape in three dimensions and relate potential deposits to prominent features. This will simplify the location of possible sources in the field. Also, as a part of this process, access to the area has to be considered. The intersections between road and rail links and the deposits are promising places to start; not only because it will be able to travel relatively easily but also because engineering works connected with infrastructure often remove weathered topsoils and expose materials of interest. Using the maps it should be possible to plan a field study to use available time and resources most effectively.

A word of warning is appropriate here. It obviously helps to have some local knowledge and it would be advisable to inform local authorities, both to enlist their help and to ensure your intentions are clear and you do not break local or national laws or regulations. Similarly, whether land is in private, corporate or co-operative ownership the rights of occupants must be respected and consents obtained. This is another good reason for restricting initial activities to road and rail links.

In the field

No general advice will ever be totally applicable to individual circumstances as each occurrence of pozzolana will be unique. The approach that will be adopted here will be to describe two actual field studies and the techniques used in preliminary and detailed evaluation of deposits.

Koru, Kenya

The Koru field study was undertaken for Intermediate Technology (ITDG) in July 1985.[1] It was preceded by a project proposal based on a desk study identifying fourteen potential sites of pozzolanic materials. The techniques employed were those described above and they were also used to prioritize the various locations so that maximum effort could be directed at the most promising materials.

Figure 1 shows the various locations in relation to a lime plant at Koru in western Kenya, while Figure 2 shows the main geological formations in the area.

In sampling materials, care was taken to discard contaminated surface deposits. The number and location of samples from each site depended on the

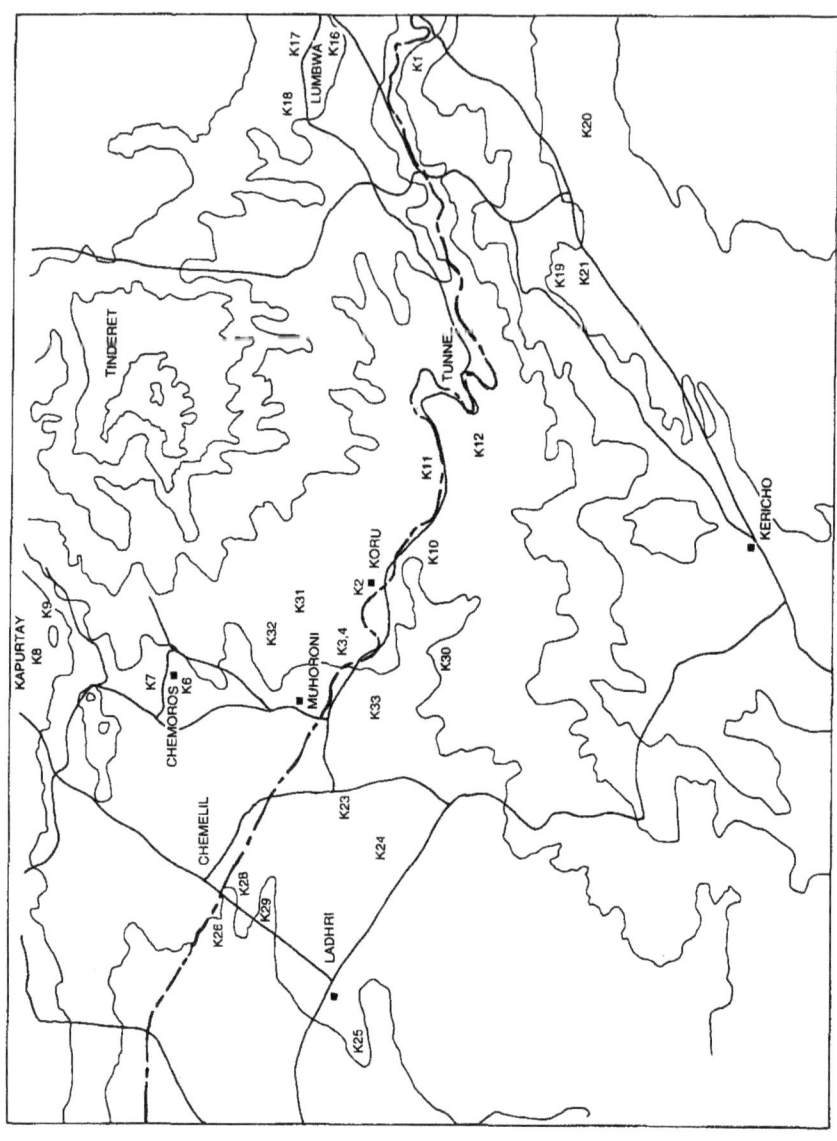

Figure 1: *Sketch map of sample sites*

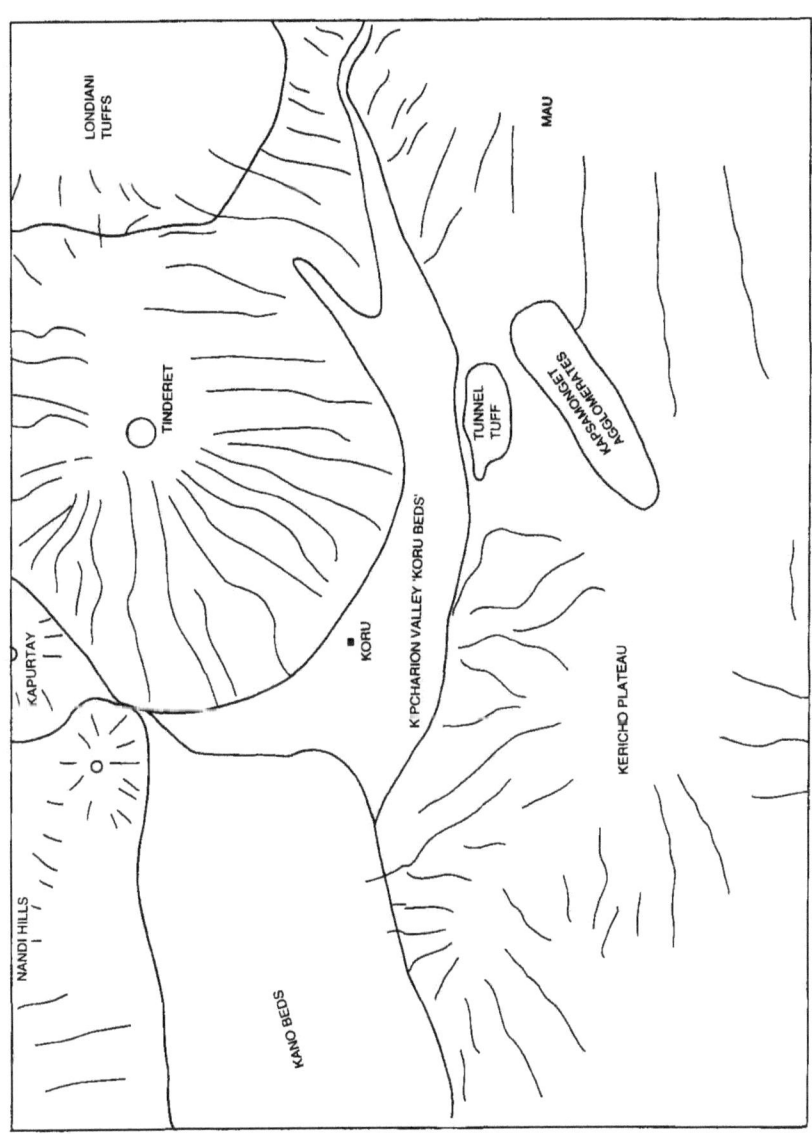

Figure 2: *Geological formation*

nature of the material *in situ*. If a vertical face was exposed and significant variations observed, several samples were taken. Careful notes were made and photographs taken to assist in the future interpretation of test results.

In general terms, close inspection with a hand lens (a very useful tool for identification) showed that darker bands in such deposits tended to contain a greater proportion of coarse crystalline material. The finer amorphous samples tended to be lighter in colour. It is also true that a lighter colour tends to indicate relatively high silica and alumina contents which are desirable in pozzolanic materials.

In this study, a rapid evaluation test was employed at Koru in order to select samples to be returned to laboratories for further testing. It was based on a method suggested by the British Geological Society, among others, and involved mixing 100cc of lime solution with 5gm of pozzolana and, using indicator strips, measuring the reduction in solution alkalinity after 24 hours.

Lime solution will have a very high pH, greater than 12. Pozzolanic materials will combine with lime, reducing the solution pH. This reduction can be easily measured using colour-graded litmus paper, giving a crude indication of pozzolanic reactivity.

A value of pH less than 10 was taken as indicating pozzolanic activity. It must be stressed that this test is only a rough guide giving qualitative results

Table 1. Test results from laboratory work at Koru

Sample	Activity	pH at 24 hours	Location	Description
K1.A	5.7	10.0	Kedowa	Welded ash
K2.A	–	10.5	Gherta (Koru)	Cemented tuff
K3.B	–	12.0	Mnara (Koru)	" "
K4.B	–	11.0	Mnara (Koru)	Weathered tuff
K6.B	–	10.0	Chemoros	Cemented tuff
K7.B	1.68	11.5	Chemoros North	Tuff
K7.C	1.62	10.5	" "	"
K8.A	–	11.0	Kapurtay	Green tuff
K8.B	3.07	7.5	"	Brown tuff
K9.A	0.37	10.0	Kapurtay North	Green tuff
K12.A	2.02	10.5	Tunnel	Welded tuff
K12.B	2.82	7.5	"	" "
K13.B	–	8.0	Tinderet Road	Welded tuff
K15.A	3.32	8.5	Kedowa North	" "
K15.B	–	10.0	" "	" "
K15.C	–	8.5	Kipkelion North	" "
K19.B	–	8.0	Kapsamonget	Pink tuff
K22	–	12.0	Ladhri Rd	Cemented tuff
K23	0.64	8.0	" "	Welded tuff
K26.A	1.15	8.5	Chemelil Station	Alluvium
K26.B	1.77	7.5	" "	Welded ash

and is no substitute for compressive strength testing. The results from an accelerated strength test[3] are given in Table 1 with the pH determinations. The poor correlation between the two is obvious.

It was clear from an inspection of the results and study notes that the most reactive materials were, as expected, the youngest (subject to least weathering), previously undisturbed, and pyroclastic in origin. They were the Londiani Tuffs, described in the literature as ignimbrites. Results were variable, even within the same formation, and this probably related to the banding noted during sampling. It appeared that sorting of amorphous and crystalline material had occurred during initial deposition.

Other materials showing promise clearly came from the Kano Beds (secondary deposits albeit recent in geological terms) having been reworked and sorted by the action of water. Reactivity was found to be variable and again significant variations in the content of crystalline and amorphous material were noted during sampling.

The study recommended further evaluation of promising materials to establish the causes of variations in reactivity and their suitability for commercial use in the manufacture of cement and building materials. Although further work was not undertaken in this case, a similar detailed investigation had been completed on the ash deposits around Oldonyo Sambu near Arusha in Northern Tanzania.

Oldonyo Sambu, Tanzania

The evaluation of a deposit of volcanic ash and pumice was undertaken in 1979. The studies were intended to assist location of reactive pozzolana for lime-based production of an alternative cement. The reactivity of the material had already been established and so the investigation focused on likely causes for variations in reactivity.

The approach adopted was to sample across the area of deposition and to correlate various material properties with compressive strength tests.[3] At the same time, consideration was given to the shape of the land and the way in which it had been formed, and to any relationship between landscape (topography) and reactivity.

The testing programme indicated that the following factors heavily influenced reactivity:

- the proportion of ash (active) to pumice and mineral content (less reactive or inert).
- surface area of the pozzolana (related to fineness).

Reactivity was also related to the topography with very interesting results. The most reactive pozzolanas were located in and around a feature described in the geological commentary as an old lake bed.

The link between the two sets of relationships lies in the way in which the landforms evolved. Study of the landscape suggests that the ash was originally deposited as a pyroclastic material (falling from the air). Subsequently this material was reworked by surface water and effectively sorted into active and inert components by virtue of the differential settling velocities of low density pumice, the ash, and the high density mineral fraction. The highly reactive ash was therefore to be found where water had concentrated the ash, in the old lake bed.

It follows that the least reactive material is also to be found in these areas and this was confirmed by a parallel study of change in reactivity with depth for samples taken from an ash face in the region of the lake bed. High pumice (at the surface) and mineral contents (at 4.0m depth) depress the pozzolanic reactivity.

Summary and conclusion

The important determinants of pozzolanic reactivity have been examined and related to the geological terminology associated with volcanic rocks. This information can be used to locate potential deposits of pozzolanic materials using geological and topographical maps.

Requirements for a preliminary field study have been reviewed and guidance given on initial evaluation of samples.

Two field studies have been described illustrating both an initial evaluation of potentially useful materials in an area of interest and a detailed investigation of a known deposit of a reactive pozzolana.

The influence on reactivity of the processes that shape the land has been demonstrated. This link can be used to assist in the location of the most reactive materials.

9. Potentialities and constraints for using pozzolanas as alternative binders in Kenya

RAFAEL TUTS*

One of the main fields of activity of the Housing Research and Development Unit (HRDU) of the University of Nairobi is the research, development and dissemination of cost-effective building materials for housing in Kenya. After concentrating mainly on walling (especially stabilized soil blocks) and roofing materials (especially fibre-concrete roofing tiles), HRDU has started up a research component focusing on binders.

---

*Housing Research and Development Unit, University of Nairobi.

The reason for this is that, for a typical low-income housing unit, the binder is the single most costly material, occurring in walls (blocks, mortar and plaster), roofs (if cement tiles are used), floors (concrete floor or stabilized floor) and foundations. Cement, the most widespread binder in Kenya, is sometimes used unnecessarily for applications where other binders could easily perform the necessary requirements. A cheaper substitute for cement would be one of the most cost-effective modifications for the low-cost bracket of the housing market in Kenya.

The importance of a new binder depends largely on the availability, variety and the relative costs of existing binders in a region, and prevailing building practice. In several countries, the development of cheaper binders has already received considerable attention from research institutes and private entrepreneurs. Different paths have been followed, but the main direction has been the (re-)introduction of lime and pozzolanic materials. The purpose is generally to increase local production of binders and reduce the cost of binders while ensuring the same or better quality of the end-product. Within the field of pozzolanic binders, the largest experience has been gathered from experiments with blended cements and lime-pozzolana binders. This direction has also been followed in Kenya, both with natural and artificial pozzolanas.

Binders in Kenya

Portland cement

Portland cement is currently produced in Kenya by two factories. East African Portland Cement (EAPC) is located at Athi River and produces a cement consisting of 80 per cent clinker, 15 per cent pozzolana and 5 per cent gypsum. The rated capacity of the factory is 350 000 tonnes per year. Bamburi Portland Cement Company (BPCC) is located near Mombasa and produces a cement consisting of 95 per cent clinker and 5 per cent gypsum. The rated capacity of the factory is 1.2 million tonnes per year.

There is a clear imbalance between the location of the cement producers and the location of consumers. About 35 per cent of the cement is consumed in areas which are between 200 and 400km from the nearest cement factory.

To minimize transport costs, EAPC would cater for the demand of Nairobi, Central, Rift Valley, Western and Nyanza provinces, representing about 81 per cent of the total market, currently estimated at 1.2 million tonnes per annum, which is impossible in view of its current capacity. A study commissioned by the Kenya Government has concluded that it is impossible to increase capacity for technical, economic and financial reasons. One reason is that the limestone quarry is located at a distance of more than 100km from the factory.

On the other hand, the BPCC factory has a clear over-capacity, in view of export problems caused by current international price levels. Therefore the surplus of BPCC is used for supplying parts of Kenya west of Nairobi. However, BPCC has experienced financial constraints for which the company is citing stiff price controls by the Government as the core of the problem. This however means that the cement has to be transported between 500 and 1000km. To prevent penalizing the distant consumers, the Government has fixed a standard price per tonne in all towns on the railway. If one considers the difference in production cost per tonne (1984) between EAPC (KSh1030) and BPCC (KSh650) it is clear that this difference covers the transportation and distribution costs of the BPCC factory.

Table 1. Consumption of cement in Kenya (1980–94) [23, 24, 25]

Year	Consumption ('000 tonnes)			Population Inhabitants	Consumption kg/inhab/year
	EAPC	BPCC	Total		
1980	293	398	691	16 700 000	41.4
1982	299	280	579	18 000 000	32.2
1984	310	236	546	19 600 000	27.9
1986	331	441	772	21 200 000	36.4
1988	322	646	968	23 000 000	42.1
1990			1178*	24 900 000**	47.3
1992			1425*	27 000 000**	52.8
1994			1724*	29 300 000**	58.8

* Extrapolations on the basis of an increase of 10% per annum.
** Assumed population increase is 4.1% per annum.

Source: Author's computations on the basis of Ref. 9.

It is clear from Table 1 that, if current trends continue, from 1993 the total rated capacity of existing plants will not be enough to cater for local consumption. On top of local production, cement now has to be imported, for instance from Egypt (1990). Since large-scale importing of cement is an option which is very difficult to defend in view of the foreign exchange involved, serious new investments are necessary. This point is also enhancing the possible role of small-scale units making binders. Their function is not only to provide lower priced cement, but also, however small it may be, to increasing the total local production of binders, and therefore freeing Portland cement for other uses.

Currently the following cement prices per tonne are applicable over the country (1991). The data are (in principle) maximum prices. It must be noted that in practice higher prices are very often noted. Table 2 shows that there is a maximum difference in cement price of 80 per cent (between Mombasa Mainland North and Mandera). The rice areas (R), which are of particular interest in this study, have cement prices which are, for

Table 2. Cost of cement in different locations in Kenya (1990–91)

Location	Cost of cement		MSA	NBI
	KSh/bag 1990	KSh/bag 1991*	=100	=100
Mombasa North	86/70	122/10 (C)	100	88
Nairobi	98/45	137/50	114	100
Eldoret	98/45	137/50	114	100
Thika	98/45	137/50	114	100
Kisumu (R)	98/45	138/80	114	100
Nyeri	100/–	139/–	115	102
Embu	102/–	141/50	118	104
Garsen (R)	108/95	150/30	126	111
Kisii	110/15	151/80	127	112
Maralal	112/60	154/90	130	114
Wanguru (Mwea) (R)	115/–	160/–	138	122
Lamu	118/35	162/20	137	120
Mandera	156/30	210/20	180	159

R Rice growing areas: Mwea/Wanguru; Kisumu/Ahero; Tana River.
* Almost 40% increase between May 1990 and May 1991.
C Currency: in October 1991, US$1 was equal to KSh 29/–.

Source: Author's computations on the basis of Refs 14, 26.

Wanguru, 15 per cent higher than Nairobi, for Kisumu the same price as Nairobi, and for Garsen 11 per cent higher than Nairobi.

An aspect which is important when considering alternatives to conventional binders is the import content of about 24.1 per cent which is required to produce one tonne of cement (Athi River). This import mainly covers equipment, spare parts and furnace oil. Any proposal of cement replacement should be evaluated at the macro-economic level in terms of expected amount of import content involved, and thus the eventual amount of foreign exchange saved.

So as to relate the cost of cement to the buying power of the consumers, and especially the low income citizens, it is interesting to compare cement prices in the capitals of different countries over the world. We tried to estimate how much a tonne of cement was costing in 1987 (and for some countries 1991), how much an unskilled worker was earning in the same year, and then how long an unskilled worker was supposed to work to be able to buy one bag of cement. In the majority of the countries, the absolute cost of a tonne of cement is varying between US$50 and US$100. The only countries which have clearly higher cement prices are Nigeria and Rwanda (which has one of the highest cement prices in the world). However, when related to income level, the countries are split into two broad categories: one with less than two hours of work to buy a bag of cement, while the second group consists of countries where more than 14 hours have to be worked in order to earn a bag of cement. On the basis of the 1991 records of Kenya and Zambia, it can be seen that the affordability of cement is dramatically worsening in some countries.

Lime

Lime, which could partly be an alternative to cement as a binder, is not extensively used in the construction sector, mainly because of cost and availability. Athough it is generally recognized that lime has a number of clear advantages (such as high workability and water retention) for some building applications, most of the lime production from existing plants is consumed in other sectors of the economy. The main applications in the building sector are as a plasticizer for plastering walls, for whitewashing and for soil stabilization in road construction.

White hydrate of lime – calcium hydroxide, $Ca(OH)_2$ – is produced by two companies. The calcium oxide content amounts to 70 per cent. Grey hydrate of lime is produced by one company. The calcium oxide content amounts to 55 per cent. These products are sold at KSh1980 (white) and KSh1584 (grey) per tonne wholesale ex-factory (1990). Besides, there are many small-scale production units of lime, concentrated along the coastal strip.

Homa Lime Company is selling limestone in bulk at KSh228 per tonne ex-Koru factory. Further, there are numerous areas where limestone can be found in Kenya, especially in the Rift Valley and along the coast.

Research methodology

On the basis of the above situation for binders in Kenya, clear constraints are noted, in terms of both availability and cost. Hence, there seems to be room to explore other avenues. According to a French mission in Kenya on low-cost binders, the production of hydraulic binders could be increased by producing pozzolanic cements or lime-pozzolana binders. They proposed three different scenarios, but all of them required an investment of more than KSh30 million.[13] Until recently, there has been no attempt to address the issue of small-scale production of binders.

The following methodology has been followed in order to determine the viability of such small-scale production units of pozzolanic binders in Kenya:

- by means of a pre-feasibility study, the need and role for alternative binders has been established
- the quantities of the possible raw materials have been estimated
- a suitable production technology has been selected and adapted to the local context
- the composition and properties of the most promising alternative binders have been determined through physical and chemical laboratory tests

- application tests are being carried out to compare selected binders with the conventional ones in terms of long-term durability
- to refine the feasibility calculations and carry out a market study for alternative binders bearing in mind the changed situation of existing binders
- to establish location and make a detailed investment and running cost plan for a pilot plant
- to assist with the installation of the pilot plant
- to set up a framework for dissemination of the technology in the relevant districts.

Rice husk ash

Global evaluation of quantities

When compared to other countries worldwide, Kenya is one of the smaller producers of rice. Kenya's rice production is only 20.4 per cent of the Tanzanian level, 3.4 per cent of the Nigerian level, which is in its turn only 1.6 per cent of India's rice production. In Africa alone, there are about 40 countries where rice is grown. Together they share an annual production of about 8.559 million tonnes/year (1981). This is comparable with the annual rice production of the USA.

Quantities of husk and ash in different regions of Kenya

The rice schemes in Kenya which are under the National Irrigation Board (NIB) are the following: Mwea in Central Province; and Ahero, West-Kano and Bunyala in Nyanza Province. There are proposals for extension of the Mwea and Ahero area and for developing the Tana Delta area for rice (near Garsen township).

Mwea Rice Millers Ltd (MRM), based in Wanguru, is a joint-venture between the National Irrigation Board (NIB) and the Mwea Farmers Co-operative. In the Mwea area there is the MRM hulling mill and a few other small mills. Around Kisumu, there are three milling companies in operation. Besides United Millers, the largest, there are also Nyando Millers (Ahero) and Kibos Millers (Kisumu).

Currently, Mwea/Wanguru is the most important source of husks. In Table 3, one can see that after the proposed extensions were implemented, the Tana Delta irrigation scheme would be the most important rice zone, followed by Mwea/Wanguru.

It should be noted that in all cases the husks are not only a by-product but also a real waste product. The husks are not used as fuel for commercial boilers, as practised sometimes in Asia.

In Mwea, the husks are brought by a MRM factory tipper to a place along the Embu road on a daily basis, where the husks are burnt when they

Table 3. Potential quantities of husk available in Kenya

Scheme	Husk (tonnes/year)				
	Existing			Proposed	Total
	Formal	Informal	Total		
Mwea/Wanguru	5600	1000	6600	4368	10968
Ahero/Kisumu	2767	300	3067	–	3067
West-Kano	400	100	500	–	500
Bunyala	200	40	240	–	240
Tana Delta	–	–	–	14560	14560
Total	6700	1440	8140	18928	27068

Source: NIB figures, field interviews, Refs 27, 6.

exceed a certain quantity. The resulting ash flies away and is washed by rain. This process does not require any fuel except for transport to bring the husks to the central burning location. This means that the dumping of husk is a real cost for the factory in terms of the maintenance of a lorry and the salary of a driver. The heat energy which is generated is not used and, in addition, air pollution is created in the process.

In 1987, some interest was shown in the husks for the purpose of possible production of 'charcoal' out of compressed husks, without any follow-up. Another investor has made a feasibility study on using the rice husks for 'rice burn oil' through solvent extraction and, further, for the production of 'Furfural' for the drug industry. The latter idea has been dismissed since it would require 15 000 tonnes of husks to be processed annually. This is clearly beyond current production in Kenya. The use of the husks for this purpose would not exclude their use as a source of ash for a binder, since only oil is extracted from them.

In the absence of feasible alternative uses, Mwea Rice Mills is interested in the binder project, and is willing to sell the husks at a nominal value. However the real economic value of the husk can only be evaluated when two or more different users are competing for the same waste product.

In Ahero, the husks are usually left on the fields, since the quantity is too little to hinder production activities. It has been observed that some people mix the burnt husks with mud while plastering their mud and wattle houses.

The potential total amount of binder produced from RHA is very low as compared to the general consumption of cement. Consider the example of a production process where a 30/70 mix of ash and OPC is used. In this case, currently 1628 tonnes of ash could be produced yearly in Kenya. This would represent only 0.47 per cent of the total cement consumption on a national basis in 1989. If we consider the proposed extension of the rice schemes, and then compare the figures with the projected consumption of 1992, about 5414 tonnes out of a total cement consumption of 1 425 000 tonnes could be produced on a national basis. This means 1.26 per cent as compared to the total consumption of cement, but a considerably larger percentage if we only consider cement used for non-reinforced applications.

When the figures are related to regional consumption, Mwea could provide 3.18 per cent of Central Province's consumption (1992), Tana Delta could provide 3.44 per cent of Coast and Eastern Province's consumption and the remaining rice areas could provide only 1.1 per cent of consumption in Western and Nyanza Province.

The relevance of considerations related to the availability of raw materials is, however, only limited to the macro economic level. If a product proves to be competitive and cost-effective, however small the production level may be, this product can be viable for a limited market.

Production procedure

In the literature, six production procedures have been reported, which could in principle be applied in Kenya.[10,11,15,17,18] These different processes of producing RHA can be schematically summarized as shown in Table 4.

Out of these possibilities the option 1B has been retained because of the following reasons:

- controlled burning increases the reactivity of the ash and does not require major capital investments; therefore heap burning (2A and 2B) is excluded.
- in order to produce lime from lime sludge and reactive clay from clay balls, higher temperatures are required than the optimum temperatures for rice husk ash; therefore lime sludge and clay ash (3A and 3B) are excluded.

Table 4. Schematic overview of known production processes of RHA

	A (lime poz.)	B (cement ext.)
1	Controlled burning Grinding RHA only Use as lime-pozzo binder	x Controlled burning x Grinding RHA only x Use as pozzo-portland binder
2	Heap burning Grinding RHA and lime Use as lime-pozzo binder	Heap burning Grinding RHA and cement Use as pozzo-portland binder
3	Burn husk and lime sludge balls Grinding resulting ash Use as lime-pozzo binder	Burn husk and clay balls Grinding resulting ash Use as pozzo-portland binder

Source: Author.

- lime for building is far less popular than cement in most parts of Kenya because of its high cost and relatively low quality for building applications; therefore lime applications (1A) were given lower preference than the use of the ash as a cement extender.

The production procedure which is retained (1B) can be described as follows. The husks are burnt in a controlled manner (less than 700°C) in a purpose-built incinerator. Control is achieved by variable air inlets. This results in an amorphous ash with a colour ranging from dark grey to white, depending on the residual carbon content of the ash. The incinerator can be constructed using brickwork and an iron sheet cover. The resulting ash is ground in a ball mill for about 5 hours to the required fineness. Three parts of cement are mixed with seven parts of the ground ash and then used as a binder, for mortars, rendering and non-reinforced concrete.

It is clear from the above descriptions that the ball mill is an important and essential part in the production equipment. The experimental ball mill of the HRDU laboratory has been designed according to the dimensions of the smallest ball mill which could be found in the literature.

Technical characteristics

Chemical analysis

In Table 5, a typical chemical composition of Asian RHA is compared with a sample analysed in Malawi and, from Kenya, samples from Mwea/Wanguru and Kisumu/Ahero. The chemical analysis of the sample from Mwea focused on the determination of the SiO_2 content.

Table 5. Chemical analysis of RHA samples (Asia, Malawi, Kenya)

Constituent	Percentage by weight (%)			
	Asian[4]	Malawi[22]	Mwea/Wanguru*	Kisumu/Ahero
SiO_2	92.15	88.30	85.00	89.44
Al_2O_3	0.41	–	0.32	0.46
Fe_2O_3	0.21	–	0.23	0.41
Mn_2O_3	–	–	0.12	0.14
CaO	0.41	0.50	0.67	0.58
MgO	0.45	–	0.43	0.42
Na_2O	0.08	–	0.07	0.47
K_2O	2.31	3.40	1.24	1.35
P_2O_5	–	2.90	0.71	1.55
Loss on ignition	2.77	2.90	6.93	3.66

* The Kenyan tests were carried out by Dr Kamau, Mr Mbindyo and Mr Kithinji of Dept of Chemistry, University of Nairobi.

The results indicate that on the basis of the chemical composition, especially the total percentage of the oxides of silica, iron and aluminium, the Kenyan RHA is potentially suitable as a pozzolanic material.

XRD analysis

XRD spectra analysis showed that, while the ignited ash indicated crystallinity, the controlled burn RHA samples were amorphous.

Fineness and grinding time

When measuring the Blaine fineness (cm^2/g) at different grinding periods it was found that the fineness (and, related to that, the crushing strength) gradually increased from 2000 Blaine to 4200 Blaine when grinding between two and five hours. When grinding over longer periods, the fineness did not significantly increase.

Crushing strength

However, the test which is most popular for estimating the real performance of the final RHA binder is the crushing strength test. The crushing strength of a test sample of one part of binder (ash and lime) and three parts of standard sand is measured after 7 and 28 days. In Table 6, the resulting strengths are compared with similar tests in the literature on RHA. It should be noted that the tests have been carried out with cement from the BPCC factory, since in the EAPC products, 10 per cent natural pozzolanic materials are already added during the production process.

Table 6. Crushing test standards and test results

Standard/Test results	Crushing strength (N/mm^2)		
	7 days	28 days	90 days
Portland Pozzolana Cement IS 1489–1967	17	–	–
Ordinary Portland Cement IS 269–1967	22	–	–
Ordinary Portland Cement BS 12:1971	23	–	–
Ordinary Portland Cement BS 12:1978	–	41	–
OPC/RHA blended cement India 1	27	42	–
OPC/RHA blended cement India 2	6	10	–
OPC/RHA blended cement Laos	21	–	–
OPC cement Kenya (control sample 100/0)	15	26	28
OPC/RHA blended cement Kenya (85/15)	16	29	31
OPC/blended cement Kenya (80/20)	15	26	30
OPC/RHA blended cement Kenya (70/30)	15	22	28
Lime/RHA (60/40)	11	–	–

Source: Refs 1, 6, 7, 8, 16, 21.

It can be observed that, for Kenya, the 28-day strength of mixtures with 15 per cent of RHA are higher than the control samples. When looking at the 90-day compressive strength, all the OPC/RHA mixtures give similar or higher results than the control sample. On the basis of the compressive strength results, it became clear that it is possible to add up to 30 per cent of RHA to the cement in a mortar mix, without losing any strength.

The compressive strength is but one specification which is normally required during the quality control of a cementitious binder. Other properties include bending strength, setting time, shrinkage and soundness.[19] These tests have until now in Kenya only been done for natural pozzolanas.

Application tests

In addition to mortar cubes, some soil blocks were manufactured with 6 per cent of 20:80 RHA/OPC mixtures as stabilizer. The resulting compressive strengths are suitable for low-cost housing applications. Further, a series of RHA plasters on soil block walls are being monitored for long-term durability. Here experiments are being undertaken with different proportions from conventional cement/sand mixes, in order to achieve the same performances. It has been observed that when utilizing up to 50 per cent of RHA as a replacement for OPC, the resulting mortars still perform very well. This would result in a materials cost saving of up to 36 per cent as compared to the classical 1:6 OPC/sand mortar (Table 7).

It should be stressed that besides the material costs of the mortars, which are compared further in this report, there are a number of other variables which could contribute to the final mortar cost per area covered. There may be differences in labour for applying the mortar and differences in the workable thickness of the mortar (e.g. for rendering).

Besides the renderings, it is suggested that RHA/OPC cement can be used as a binder for mortars, floors and fibre-cement tiles and applica-

Table 7. Cost benefits of alternative mortars for wall rendering

No.	Quantities by weight (g) of inputs					Cost KSh/m^2	% No.6 = 100%
	RHA	OPC	LIME	SOIL	SAND		
1.	500	500	0	0	6000	13.3	77%
2.	500	500	0	2000	4000	12.9	64%
3.	300	700	0	2000	4000	17.9	73%
4.	300	700	0	0	6000	14.9	86%
5.	300	350	350	0	6000	15.1	78%
6.	0	1000	0	0	6000	25.5	100%
7.	0	1000	0	2000	4000	21.4	87%

Source: HRDU test results.

tions other than reinforced concrete. The performance comparisons can of course only be carried out after an elaborate programme of application and field tests.

Standards and specifications

Comprehensive standard specifications for any pozzolanic binder on the basis of incineration of plant wastes should include:

- methods of preparing the binder
- chemical and physical properties of the pozzolana
- characteristics of the pozzolanic binder
- storage conditions
- methods of use
- curing conditions.

Economic feasibility

In order to check the economic feasibility of the most interesting options (in terms of combination of location of the plant and the suitable sources of raw materials) we have developed a spreadsheet program which is calculating several economic feasibility criteria on the basis of the relevant parameters. A number of parameters have been kept constant during the study.

An economic feasibility analysis has been applied to the three locations for the selected production processes. In Table 8 we give a summary of the results. For each case, the following four characteristics have been listed:

- production level (tonnes/day)
- amount of the yearly available husks utilized (%)
- cost of the product (KSh/tonne)
- cost of a mortar based on the binder as compared to a cement mortar (%).

Table 8. Summary of economic feasibility for the different areas

Area	Mix	Cost Cement KSh/t	Cost Lime KSh/t	Tonne/ Day	%	Cost RHA KSh/t	Mortar cost Cement KSh/t	Lime %	RHA %
Mwea	Lime Poz.	2200	2040	5	90	1379	897	104	74
	Cement Ext.			3	80	783			79
Kisumu	Lime Poz.	2100	1703	1	90	1921	873	101	90
	Cement Ext.			0.75	100	1909			93
Tana	Lime Poz.	2400	1044	11.25	80	809	946	60	55
	Cement Ext.			7.5	80	567			88

Source: Author.

It should be noted that for the last columns, mortar prices are compared. Since the sales price of binders are usually expressed by weight (KSh/tonne), while the batching is usually done by volume, densities (kg/m³) are necessary for converting bulk prices to mortar prices.

While the previous calculations of the Mwea case were based on a relatively independent production unit, other assumptions, related to the organizational structure of the plant, can also be made. A first modified set of assumptions simulates a unit completely built in the MRM factory, which would require less overheads. This means that the supervisor can be omitted, the office space and the site are assumed to be part of MRM properties, implementation costs can be taken over by MRM and there is no need for a loan. In this case the economic characteristics would be changed favourably and lead to a reduction of the RHA cost by 12 per cent. Other, less favourable assumptions, simulating a plant which would be more independent from MRM can also be imagined. Here the need arises for purchasing a pick-up or a tipper, and the cost of the final product could be increased by up to 21 per cent. From this, one can see how the status of the production unit can determine the level of investment and how this will influence the marketability of the product.

In Table 9, some other parameters are checked on their sensitivity to influence the final cost of the product. The final price of the product per tonne can be read against the varying costs of the inputs. One can notice that the cost of husks is a very sensitive parameter, while the cost of labour can only influence the selling price of the product marginally.

Another factor which needs to be analysed is the distance between the consumer and the production unit. In view of prevailing transport costs, it is clear that, in most of the options, the product will be out of the market beyond a radius of 40km. This does not pose a problem, since the limited capacity of the units simply does not allow for sales beyond a similar radius from the production site.

Table 9. Sensitivity analysis of cost of lime, husks and labour

Lime cost KSh/tonne		Working days No. days		Labour cost KSh/day		Husk cost KSh/tonne	
2100	1488					300	2322
2000	1455					200	1988
1900	1422					100	1655
1800	1388			80	1405	50	1488
1700	1355	350	1281	70	1364	20	1388
1600	1322	300	1322	60	1322	0	1322
1500	1288	250	1379	50	1280		
1400	1255	200	1466	40	1238		
1300	1222	150	1610				
1200	1188	100	1899				

Natural pozzolanas

Athi River Cement Company adds about 10 per cent of a pinkish tuff, quarried near Athi River, to its cement. KS02–21:1976 allows a maximum of 15 per cent of natural pozzolana as additive in factory-produced cement. In Kenya, natural pozzolanic materials are abundant. The main types of pozzolanas are pumices, diatomites, phonolites and tuffs. They are lying largely unexploited.

A selection of these materials has been tested[5] so as to ascertain their pozzolanic properties. The materials tested were: pumice from Longonot; phonolite from Embakasi (Nairobi); diatomite from Kariandusi (Gilgil); and Yellow tuff and Athi tuff from Athi River. These materials were found suitable as pozzolanas to varying degrees (50 to 15 per cent) of cement replacement without adversely affecting important concrete properties such as strength, workability and setting time. Besides these sources of pozzolanas, HRDU has initiated tests with other natural pozzolanic materials found in the areas of Isiolo and Mt Kulal.

Table 10 illustrates the variation in qualities such as colour, reserve estimates, and reactivity of some of the potential natural pozzolanas.

Table 10. Description of selected natural pozzolanas[5]

Name	Location	Description	Colour	Reserves million tonnes
Pumice	Longonot	Small pebbles	Grey	62.4
Diatomite	Kariandusi	Soft rock	White	2.0
Phonolite	Embakasi	Dust (vumbi)	Grey	5.1
Yellow tuff	Njiri	Soft stone	Yellow	2.5
Athi tuff	Athi River	Soft stone	Pink/grey	23.0

In Table 11, some selected characteristics of the natural pozzolanas are listed. The combined values of $SiO_2+Al_2O_3+Fe_2O_3$ vary between 53 and 84 per cent. The concrete cubes strength, as compared to OPC only as binder, shows that three of the five pozzolanas have systematically higher strengths than OPC when mixed in a 15:85 ratio. For the 30:70 ratio, still one sample gets higher values than OPC. The other pozzolanas reach between 82 per cent and 89 per cent of the control value. The diatomite sample could only be tested up to 25 per cent because, with richer diatomite mixes, the water requirement becomes very high. It can also be seen that the relative strength of all the pozzolana samples (as compared to the control sample) is higher at 60 days than at 28 days, which confirms the characteristic long-term strength increase of pozzolanic binders.

No tests have yet been carried out with natural pozzolanas on the basis of lime. This could be done with lime from different areas such as Koru, Homa Bay, Ortum-Sebit, Kajiado, and the coastal areas. Since lime from

Table 11. Characteristics of selected natural pozzolanas

Name	Chemical analysis $SiO_2+Al_2O_3+Fe_2O_3$	Density (ground) kg/m³	Concrete cubes compr. strength OPC only = 100%			
			15% PZ		30% PZ	
			28 d	60 d	28 d	60 d
Pumice	82%	2475	92%	94%	79%	83%
Diatomite	84%	2020	109%	114%	–	–
Phonolite	67%	2705	88%	93%	77%	82%
Yellow tuff	53%	2240	120%	124%	102%	106%
Athi tuff	76%	2430	100%	102%	82%	89%

Source: Author's calculations on the basis on Ref. 5.

these areas is not at all standardized and is very variable in quality, such a test programme would have to take into account a much larger number of variables than when using cement as the source of calcium hydroxide.

Conclusions

General potential

The development of RHA binders can only be of interest in countries or regions where: (a) rice is produced in sufficient quantities, where (b) OPC is relatively expensive and where (c) there is a need for an increased stock of binder. These basic conditions for successful applications of RHA binders are met in the three rice growing areas of Kenya, but not in a convincing manner.

This is so because (a) the quantities of rice produced are rather limited, (b) the price of OPC is controlled by Government and kept at a relatively low level and (c) the population growth is so high that the increased demand for cementitious binders cannot be met by RHA potential alone.

Rice husk availability

Currently, there is no viable alternative use for the rice husks. This is applicable for Mwea and Western rice areas; for Tana Delta it is of course not yet known whether the husks will be used for any purpose other than RHA (possible uses are cattle fodder, fuel and manure) but the possibility of starting rice husk ash production from the start of the rice scheme offers potential for optimal use of the husk as a binder in the Tana Delta. The value of the husks is so low that the factory must be located within, or as close as possible to, the grinding mill, in order to save expenses on transport.

Equipment

In terms of equipment, the production of ball mills does not give problems in Kenya; several producers were interested in the idea and could quote

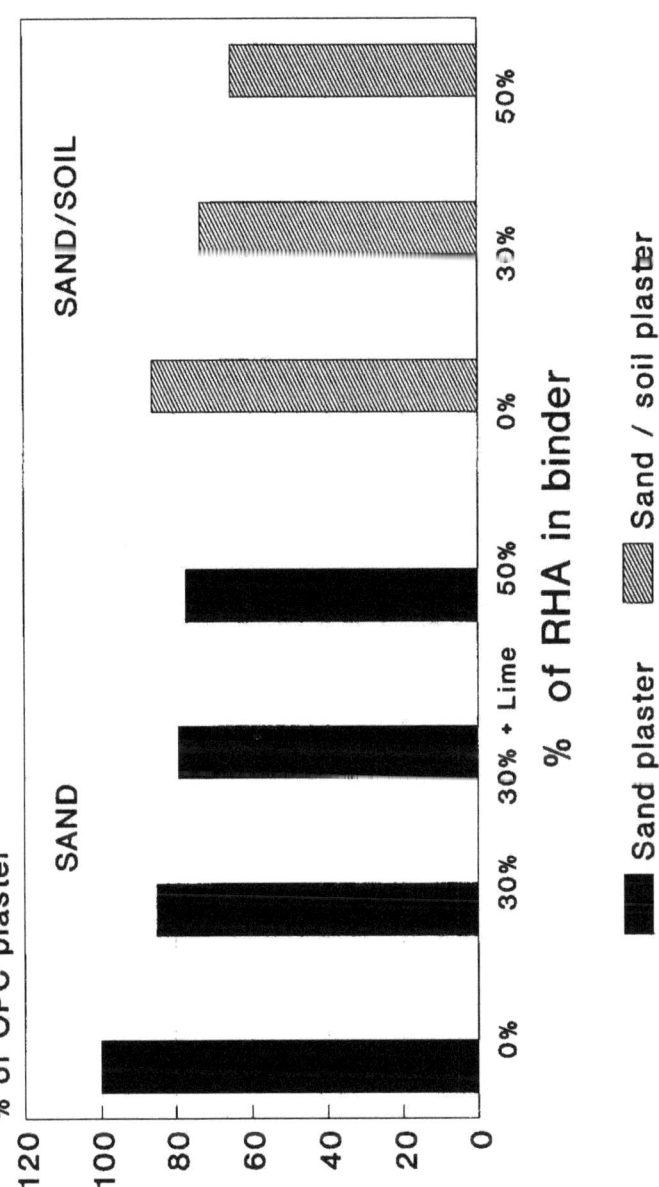

Figure 1: *Cost comparison of plasters including rice husk ash, ordinary Portland cement, lime, soil and sand*

reasonable prices for the proposed equipment. Attention should be paid to the possibility of recovering the fuel value of the husks.

Choice of production process

One of the most critical factors is the availability, cost and quality of lime. This will largely determine whether to go for a blended cement type of RHA binder or to go for a lime-pozzolana RHA binder.

The process with controlled burning and lime has a high potential for Tana Delta (because of the low cost of locally produced lime), a limited potential for Mwea/Wanguru (because of high cost of lime to be transported from Koru) and very limited potential for Kisumu/Ahero (because of the high cost of lime and low production capacity due to small quantities of rice husk).

The process with controlled burning and use as a blended cement has a high potential for Mwea/Wanguru, limited potential for Kisumu/Ahero (because of the low production capacities envisaged) and very limited potential for Tana Delta (because of popularity of the existing lime binders).

Feasibility of the production units

The upper limit capacity of the production units (because of the limited availability of husks) is an important blocking point in the whole exercise, because production units of 5 to 10 tonnes a day would be able to produce at definitely lower cost than the proposed units of 1 to 5 tonnes a day.

Cost of the final product

It has been stressed that the final costing should be evaluated in comparison with mortars. The results of this evaluation show that for Mwea/Wanguru, the controlled burning option with lime would give a cost reduction of 26 per cent (as compared to the cost of the conventional cement mortar). In Kisumu/Ahero the cost of the controlled option would give a cost reduction of 10 per cent (as compared to the conventional cement mortar). In Tana Delta, the process of controlled burning with lime would result in a cost reduction of 45 per cent of the conventional masonry mortars.

Quality of the final product

After laboratory tests, the technical aspects comply with standards set in other countries. Application tests and field tests continue to evaluate the durability and other long-term performances of the mortars.

Macro-economic benefits

Besides the possible advantages at the consumer level, there will be definite macro-economic benefits related to the production of RHA binders.

While producing Portland cement requires about 24 per cent foreign exchange (mainly equipment and fuel), it is expected that a RHA production unit will only need 2 per cent of foreign exchange (in terms of the imported raw materials for the local manufacturing of the equipment). Further, there is a lower capital use and a higher labour use per annual tonne, and a lower capital required for creating new employment.

Acceptance and marketability

One of the most critical questions remaining, is whether the expected cost reductions will be significant enough to minimize hesitation regarding the use of RHA mortars. This factor of natural resistance against any new building product is very important.

Other types of pozzolanic materials

It is strongly advised to pursue the possibilities of utilizing natural pozzolana (volcanic ash, tuffs, scoria) and other artificial pozzolana (black cotton clays) for blended cements and lime-pozzolana binders. When following this line one would expect fewer constraints in terms of availability of raw materials, but more problems in terms of quality of the final product because of the enormous variation of reactivity of natural pozzolanas available in Kenya.

10. Rice husk ash cement
R G SMITH*

Volcanic ash mixed with lime was used for making mortar at least 2000 years ago. Powdered tiles or pottery and lime have a similar history. The use of these pozzolans with lime has continued through the ages. More recently, in the 19th century, ordinary Portland cement (OPC) was developed, and pozzolans have been added to OPC to make pozzolanic cements. For example, pulverized fuel ash, a residue from modern coal-fired electricity generation, has been mixed with cement. These products have been the subject of development over an extended period, and have been widely and successfully used.

Another type of pozzolan is the ash remaining after the husks from rice plants have been burned. The rice husk ash (RHA) is first ground to a fine powder, then mixed with either lime or OPC to give rice husk ash cement (RHAC).

---

*Building Research Establishment, Garston, UK.

Table 1. World production of rice paddy[1]

Year	Millions of tonnes	Year	Millions of tonnes	Year	Millions of tonnes
1961	216	1971	318	1981	412
1962	227	1972	308	1982	424
1963	247	1973	336	1983	452
1964	263	1974	333	1984	469
1965	254	1975	359	1985	472
1966	261	1976	350	1986	471
1967	278	1977	372	1987	465
1968	289	1978	388	1988	491
1969	296	1979	377	1989	518
1970	317	1980	399	1990	518

Availability of raw materials

Although the quantity of rice produced in a few countries has decreased recently, as other more profitable crops have been grown, the quantity in other countries has gone up by a greater amount, so that production worldwide has increased steadily. Table 1 shows world production since 1961.

The weight of husk remaining after the milling process is approximately 20 per cent of the weight of paddy. Hence, 100 million tonnes of husk was produced worldwide in 1990.

The large piles of husk which accumulate adjacent to rice mills are a nuisance, and constitute a major disposal problem. A small proportion of husk waste from mills does find a use already, for chicken litter, fuel, or for making up and stabilizing ground. However, most of the husk is a waste material. Often the pile is set alight to reduce the volume of waste. Usually the pile is near the mill and, if alight, it is a fire hazard to the mill itself. Sometimes the husk is carried to a remote corner of the site for burning.

When husk is burned, the weight of ash remaining is approximately 20 per cent of the weight of husk. Hence 20 million tonnes of ash would have resulted had all the husk been burnt in 1990.

Limestone is one of the most common minerals in the world. From it, lime can be made, by a wide variety of technologies, from simple labour-intensive low capital cost, to the other extreme of large-scale mechanized production. The extent of the potential for production of RHAC from ash and lime, worldwide, is 30 million tonnes per year (assuming proportions of 2 to 1 by weight respectively).

OPC is produced in most countries of the world, but for many housing schemes in developing countries, it is in short supply, or too expensive. The addition of RHA to the mix could allow either less expenditure on OPC for a certain amount of building construction or more building to be done with a particular quantity of OPC.

Hough and Barr examined and reported on the addition of RHA for making a cement in 1956[2], and referred to the successful making of building blocks using RHA and cement as early as 1923. In 1971, Mehta[3] published his ideas of RHAC manufacture. A workshop at Peshawar, Pakistan, brought together interested parties in 1979 for presentation of their findings[4], and in 1984 Smith reported on progress in the Indian sub-continent.[5] A comparison between RHAC and other cementitious materials together with a brief review of RHAC worldwide was presented at an international conference in Roorkee, India in 1984; this paper[6] contained references to many individual items of investigation around the world. Then in 1985 a book was published by the United Nations[7], compiling knowledge gathered by Cook from many sources.

The principle of the method of manufacture

Ash remaining after rice husks have been burnt consists very largely of silica. The chief impurity is carbon, especially prevalent if the organic constituents have not had sufficient time or sufficient supply of air to enable them to burn off completely. There will be small amounts of sodium and potassium compounds, and probably some phosphorus and other trace elements derived from the soil.

Silica can have a variety of forms, some more suitable than others for making RHAC. Crystalline forms of silica have the constituent atoms arranged in an orderly manner within the molecule and these forms are not very reactive with lime. Amorphous (non-crystalline) silica has a disordered array of atoms, with higher reactivity, so it is the preferred form. The crystalline form, crystobalite, tends to develop if the temperature is maintained above approximately 750°C, and if the temperature rises to 850–900°C, tridymite may be found. Although these crystalline forms may have some pozzolanic reactivity with lime, especially if they are very finely ground, it is preferable to limit the temperature of incineration to 750°C, so producing the more reactive amorphous silica.

Where suitable laboratory facilities exist, the ash can be sent for chemical analysis, and determination of the degree of its crystallinity, using X-ray diffraction. The latter will identify the type of any crystalline forms present, and give an indication of whether there is only a trace, moderate or abundant amount of them in the ash. Other practical tests can be carried out to assess the degree of pozzolanicity of an ash: the lime removed from OPC by the pozzolan and the amount of silica which goes into solution can give an indication.[8]

Whether or not the silica of the RHA is crystalline, it is always necessary to reduce it to a very fine powder, so that a large surface area is available for reaction with added lime. A ball mill is a suitable device for powdering the ash, though other devices have been tried.

Experimental mixes of RHA with various proportions of hydrated lime will enable the optimum to be determined. Similarly, mixes of RHA and OPC should be made. Each type of RHAC should be mixed with sand in various proportions, stored under controlled conditions and then the mechanical properties determined in the laboratory, to ascertain the properties before venturing into commercial production of RHAC. Compressive strength is an important property to measure in relation to construction and also, in general, if the compressive strength of a material (after soaking in water for 24 hours) is high, then its durability is likely to be good. Conversely, low wet compressive strength may imply poor durability. Weathering tests should be set up to substantiate durability of RHAC.

There is no national or international standard specifically for RHAC, though guidelines may be derived from standards for similar materials.[9]

Recent development work

The following data are derived from work where BRE has been involved in visits to developing countries and in joint research projects with other institutes. An outline of the background to the work is given first for those countries.

Suriname

More rice is produced per head of population in Suriname than in any other country. Rice is a major crop for local consumption and for export, and the country has a very small population. The rice is all milled within Suriname, so much husk is available, and the potential for rice husk ash cement is very great. This is especially important in Suriname, since the raw materials for manufacture of either lime or OPC do not occur there.

Waste husk is usually blown out of a pipe from the rice mill, whence it deposits in a large conical heap which is burning continuously, as it is deposited, to minimize the volume of waste. The temperature of combustion is not subject to any form of control in the heap, but probing with a thermocouple revealed that the temperature in the heap increased with distance from the outer surface up to a maximum of only 550°C just 60mm beneath the surface. Thereafter, the temperature was progressively lower with distance from the surface, up to 600mm, which was the maximum distance of reach for the thermocouple.[10]

The Suriname Ministry for Social Affairs and Housing, in conjunction with the University of Suriname, has made RHAC from this ash by grinding it in a locally made ball mill, and then mixing it with the imported OPC.

Malawi

A steel wire basket incinerator, modified from the Tube-in-Basket design of Kapur[11], was constructed at Blantyre Polytechnic, University of Malawi.

Several thermocouples were placed to record temperatures in the burning husk, and short-period peaks up to 800°C were noted. Subsequent tests showed the ash to be good quality, but even with topping up the incinerator with more husk during the burning process, the total weight of ash which could be produced in one batch in the wire basket was only 2.5kg. The cost of the incinerator was high in relation to the weight of ash produced, because steel, which is imported, is expensive in Malawi. Therefore two alternative incinerators were built of cheap locally-made bricks, laid without mortar. The first incinerator allowed bricks to be removed from a few points in the side walls to increase access of air; the second had a fixed amount of ventilation, given by vertical gaps 10mm wide between every brick.[12] The volume of the brick-built incinerators was several times that of the wire basket, yet the cost was only one-tenth. Temperatures attained in the brick-built incinerators were similar to those in the wire basket incinerator.

A small laboratory ball mill was used for grinding the ash for various lengths of time. A pestle and mortar and a maize mill were tried as alternative methods of reducing the burnt husks to powder. A sample of rice straw was burnt to investigate its properties in comparison with those of husks.

Sierra Leone

Research has been started at the Fourah Bay College of the University of Sierra Leone. A simple version of the brick-built incinerator developed in Malawi was constructed, thermocouples placed, and husk dust from a small mill in Freetown and husk from a mill in the town of Kambia were burnt separately. The husk dust was too fine to allow adequate access for air for combustion, so burning was difficult to carry out. Moreover, the husk dust contained broken rice grains, which ought more properly to be utilized for their food value, for example as chicken feed. Therefore the dust cannot be recommended as a source for pozzolan, but the Kambia husk could have potential use.

Guyana

In terms of rice produced per head of population, Guyana is second only to Suriname. Raw materials for production of lime or OPC do not exist and so, as with Suriname, the need and potential for RHAC is very great. The Institute of Applied Science and Technology has built experimental incinerators of brick with some wire mesh at the bottom to allow access of air from underneath, the latest large enough to provide half a tonne of RHA per day, sufficient to justify commercial production of RHAC.[13] Roofs over the incinerators facilitate operation, even during short rainy periods. Thermocouples within the burning husk indicated maximum temperatures of 550 and 690°C. Light-coloured ash is deposited at the bottom of the incinerator.

Laboratory examination of one piece of such ash shows it to have an open cellular structure beneath a thin skin so that, on crushing the ash, many small particles are produced.

The Institute is installing a large ball mill so a complete RHAC production plant of commercial scale can operate, producing one tonne of RHAC per day, and giving information on technical features and actual economics of production.

India

Several institutes have developed RHAC technology. The Cement Research Institute (CRI), New Delhi, developed a brick-built incinerator with a wire mesh lining, and applied its methods through approaching entrepreneurs and provided a technical backup service. The Indian Institute of Technology (IIT), Kanpur, developed the Tube-in-Basket wire mesh incinerator and experimented with grinding times required with ashes of varying degrees of crystallinity, concluding that even crystalline ash could be used if it were milled for several hours though, of course, amorphous ash was preferable. IIT acted in a consultancy role, and a plant was set up to produce RHAC commercially. The Birla Institute of Scientific Research, Ranchi investigated the addition of other wastes from another local industry, and initiated a single production plant on a neighbouring estate of small-scale production industries. The Central Building Research Institute (CBRI), Roorkee, researched the use of husks mixed with either the waste lime sludge from the local sugar industry, or clay, and developed special methods for burning these slightly different materials.

Ash-lime and ash-OPC materials were both produced commercially in India.

Nepal

The Research Centre for Applied Science and Technology, Kathmandu carried out some research which stimulated local production.

One group of local people arranged to collect the ash remaining from the burning of husk to roast rice grains for sale to the public. The ash was taken by wheelbarrow to the RHAC production plant. During a few days of storage, it changed in colour from dark grey to white, and was then milled, mixed with lime, bagged, and sold at the roadside.

Another entrepreneur attempted to increase the scale of production to meet the local demand by building a large incinerator, with very little ventilation. Inspection of the plant indicated that a very high temperature was reached, with the likelihood of producing an unreactive material.

Thailand

The Asian Institute of Technology (AIT), near Bangkok, developed medium-size incinerators, holding the ash in a wire basket, surrounded by metal sheeting to exclude rain and excess wind; while a gap between basket and sheeting allowed adequate ingress of air for combustion. As an alternative to balls in their ball mill, AIT used rods, which are cheaper, yet performed satisfactorily.

The University of Khon Kaen burnt husk in a wire mesh and brick incinerator.

Malaysia

A well-insulated metal sheet incinerator ventilated through some vertical tubes with small holes drilled in them, running up through the husk, was designed and built by the Universiti Sains Malaysia (USM) in Penang. The rate of burning was very slow, and the temperature, which was monitored with thermocouples at three points, rose to maxima of only 560, 430 and 485°C at these points. The university had a ball mill built locally.

Kenya

A wire mesh and brick incinerator was constructed and used by the Housing Research and Development Unit (HRDU) at the University of Nairobi. HRDU had a ball mill built locally.

Pakistan

The Pakistan Council for Scientific and Industrial Research, Karachi was involved in research into RHAC,[4] incorporating it within a larger project on low-cost house construction. They built a large ball mill locally, to produce the larger quantities required.

China

A sample of husk from China was burnt in the laboratory, and the ash analysed.

Laboratory tests on ash

In most of the research and development programmes mentioned above, the product was assessed either locally or in the UK by the Building Research Establishment. Some of the results are reported and compared below.

The silica in RHA is the component which may react, as a pozzolan, with lime, so a high silica content is desirable. Therefore the ash remaining after rice husks (and the rice straw from Malawi) had been burnt was analysed for silica content at BRE. Amounts shown in Table 2 are not replicate determinations; each result is for an entirely different source of ash.

Table 2. Silica content in ashes from various countries

Source country	Silica in ash (SiO_2%)					
Suriname	90	88	94	92	85	90
Malawi	87	84	65 (rice straw)			
Sierra Leone	72	89				
Guyana	86	87	74	80	80	
India & Nepal	n	n				
Thailand	86	90				
Malaysia	78	72				
China	99					

None of the Suriname ashes were from a purpose-built incinerator: the first and last in the table were from a large boiler and a small combustion device, each fired with husk, where the prime purpose was to utilize the energy content of the husk. The other four ash samples were collected from heaps burning by rice mills. Silica content is high in all six samples, though not necessarily with a high pozzolanic reaction. Chemical analysis confirmed that carbon content was low in all six samples.

The first of the Malawi RHA samples was that burnt in the wire basket; there was only 1 per cent of carbon, presumably due to the ready access for air, and the silica was quite high. The second sample was from one of the incinerators built of brick, an incinerator typical of that used in several other countries, and the ash had less than 1 per cent carbon and a fairly high content of silica. However the rice straw ash, with only 2 per cent carbon, had a very low silica content, so would be unlikely to serve as a useful additive to lime or OPC.

The first of the Sierra Leone samples was ash from dust collected from small mills in Freetown, and was low in silica, reinforcing the comment above that its use is not appropriate in cement production. The other ash, from Kambia, was typical of that produced by the milling process, the pieces of husk being some 6mm long by approximately 2mm, and with a curved form: these burnt satisfactorily, and the silica content was high.

The first of the Guyana samples was from a purpose-built incinerator and silica is high. The second two were from a very large burning heap of husk which had been pushed from the mill by bulldozer, and the variation in silica between the two is likely to be due to the non-uniform burning conditions. These two samples were coloured lilac and pink respectively, an indication of high temperatures of burning, and the likelihood of very crystalline ash with low pozzolanicity.[7] The last two samples were taken from a burning heap at a different mill, were light grey in colour and, though lower in silica, they may be more useful as pozzolans.

The first of the Thai samples from AIT is high in silica, and the second from Khon Kaen is only slightly less. Carbon was low at 2 per cent in each.

Malaysian ash burnt in the incinerator gave the first result listed above.

The fairly low silica content of only 78 per cent is explained partly by the presence of 5 per cent of carbon, which in turn may be due to a restricted access for air caused by holes in the ventilation chimneys being small. The second sample was a black coloured ash from a commercial combuster used for generating hot air for drying paddy prior to milling, and it has an even lower silica content; it had an exceptionally high carbon content of 22 per cent and is unlikely to be suitable for use in making a cement.

The Chinese ash was burnt in the laboratory, as a small quantity, having therefore relatively great access to air. The result is included to indicate the high silica content which can be obtained in some cases.

Degree of crystallinity of the ash

X-ray diffraction analysis equipment at BRE has been used to identify the nature of any crystalline components in the above mentioned samples of RHA. Results are given in Table 3.

The Suriname ashes burned in heaps are either entirely amorphous or contain only traces of crystalline material, so appear promising for cement production. Low levels of crystallinity of ash from the heating devices (first and last in the table) would indicate that they might be usable.

Ash burned in Malawi in the wire basket appears to be eminently suitable for cement production, and that from the brick-built incinerator would seem to be suitable.

Table 3. Crystalline silica phases in ashes from various countries

Source country	Phases identified in ash					
Suriname	c	*A*	ct	*A*	Act	ct
Malawi	*A*	Ac	–			
Sierra Leone	Q	q				
Guyana	q	cT	cT	Qct	Qct	
India	Aq	Aq	cryst			
Nepal	Aq	T				
Thailand	n	n				
Malaysia	Aq	C				
China	n					

Key: *A* = entirely amorphous material, no crystalline phases
 A = mainly amorphous material
 Q = moderate amount of quartz present
 q = trace of quartz detected
 C = moderate amount of cristobalite present
 c = trace of cristobalite present
 T = moderate amount of tridymite present
 t = trace of tridymite detected
 cryst = crystalline material present, not identified
 n = not assessed

Note: The samples in Table 3 are in the same relative position as those in Table 2.

The husk dust from Sierra Leone, identified above as unsuitable, does have a substantial amount of crystalline material in the low proportion of silica in this ash. In the other ash, from Kambia, there is only a trace of crystalline quartz, so the material might well prove satisfactory.

The two ashes from Guyana, coloured lilac and pink, and predicted to be unsuitable for cement making, do indeed contain much crystalline material, especially tridymite which forms at a higher temperature than cristobalite, thus supporting the prediction. The ash from the purpose-built incinerator, at IAST, contains no more than a trace of crystalline quartz, so would appear to be a good material for cement production. The two ashes taken from a heap at another mill (last two on the Guyana line in the table) are somewhat crystalline, though not so much as at the first on that line and, if production were contemplated at that mill, it would probably be wise to construct a special incinerator, such as that at IAST.

The Indian samples in Table 3, for which no silica content was given, are as follows: the first was the commercial plant, based on CRI technology, seen to be under careful control; the second ash from the small Tube-in-Basket device at the IIT laboratories, the third from a commercial plant advised by IIT staff. The first two ashes, being mainly amorphous, can be regarded as good, but the third is not so. During a visit to the third site it was observed that husk was moved from the vicinity of the mill and burnt in a large heap, probably in a very uncontrolled manner so, in some places at least, high temperatures might prevail, favouring formation of crystalline ash.

The first of the Nepalese samples in Table 3 was from the small rice roasting units where, because of ready cooling conditions, the temperature of combustion would not be able to rise to a high value, precluding formation of crystalline phases of silica. The second is from the over-size incinerator which was conducive to high temperatures of combustion and formation of crystalline phases. This plant design cannot be recommended.

The ash from the special incinerator at USM in Malaysia is very amorphous, with only a trace of crystalline quartz, and should be suitable for cement making. The black ash from the combuster contains cristobalite and, together with its 22 per cent carbon content, it is not likely to be suitable for cement production.

Degree of pozzolanicity of the ash

Ashes were subjected to the BS 4550 test for pozzolanicity. Results are collected together in Table 4.

Of the six ashes from Suriname tested, the two which were wholly amorphous are both highly pozzolanic, while the other two heap-burned ashes that had traces of crystalline material in them can be described as pozzolanic. The ashes from the heat-producing boilers and combusters are the least pozzolanic.

Table 4. Pozzolanicity of ashes from various countries

Source country	Phases identified in ash					
Suriname	LP	HP	P	HP	P	LP
Malawi	n	HP	LP (rice straw)			
Sierra Leone	P	P				
Guyana	HP	NP	NP	P	P	
India & Nepal	n					
Thailand	n					
Malaysia	P	P				
China	n					

Key: HP = highly pozzolanic
  P = pozzolanic
  LP = low pozzolanicity
  NP = not pozzolanic
  n = not tested

RHA from a brick-built incinerator in Malawi was found to be highly pozzolanic. The rice straw ash was barely pozzolanic, so unlikely to be of use for cement making.

Material burnt in Sierra Leone has potential for cement making, as it is pozzolanic though, as mentioned above, the first sample shown (rice dust) is not recommended for other reasons.

Of the several RHA samples from Guyana, the lilac and pink ash from the burning heap at one mill are not pozzolanic, confirming that they would not be suitable for cement making. However samples from the burning heap at another mill were pozzolanic, and could be suitable for making cement. Nevertheless, the husk burnt in a special incinerator (first listed in the table) is superior, being highly pozzolanic.

The two samples from Malaysia are both pozzolanic. The first, from the purpose-built incinerator, should be suitable for cement production but the high carbon content of the second may preclude its use, as high carbon will not only produce a very dark cement which is unlikely to be acceptable in the market, but it may also interfere with the setting properties of the cement.

Processing of ash into cement

RHA must be ground to a fine powder of similar fineness to that commonly known for OPC. Laboratory methods exist for measuring the degree of fineness, but facilities have not been available for determining this during the present studies. However, the use of a ball mill has been the norm in all the work above.

Research in Malawi also investigated the effect of milling time. Generally, although longer periods of milling resulted in greater strengths, 30 minutes was found to be a sufficient period for the type of ash produced there, bearing in mind that longer periods incur greater fuel costs, and for a given volume of production, larger and therefore more expensive mills are required. The data in this paper for Malawi RHA are all for 30-minute milling periods.

Laboratory mills holding only approximately one litre have been used in some places, while for commercial scale developments large mills have been built. One of the largest is that provided for work in Guyana, which has a volume of 2700 litres, sufficient for producing half a tonne of RHA per day.

Alternative methods of reducing the burnt husks to powder were tried, but neither pestle and mortar nor maize mill were as good as the ball mill.

OPC is available in many countries, generally produced to accepted national or international standards. Lime may be available locally, but its quality may be variable. If none is available, production from a source of limestone should be considered. A fairly simple kiln can be constructed for producing lime.

Mixing of ash and lime or OPC is best performed with the dry materials, in a blending machine although, for convenience in the development phases, to avoid buying the separate blending machine, the ball mill can be used for mixing. In commercial production, it would be more economical to use separate mill and blender, as a smaller mill would then be required for a given rate of production. Use of the mill for intermixing RHA and OPC may reduce the particle size of the OPC, this action of itself possibly increasing slightly the rate of set of the OPC itself.

Proportions of RHA, lime and OPC should be investigated in the laboratory, before decisions are taken on equipment required for commercial production. The following section records the properties which have been found with various proportions of components, in several locations.

Strength

For comparative purposes the compressive strength of mortar made from RHAC and sand, in the proportions 1:3 respectively, are quoted. The strength tests have not all been done by identical methods, and they have been carried out in several laboratories across the world. The basis of all the tests has been to prepare small cubes, and then test some after after 7 days and others after 28 days of curing under moist conditions.

Suriname

RHA from heaps of ash burnt at rice mills was mixed with various proportions of imported OPC: results are given in Table 5.

Table 5. Compressive strength of Suriname mortar cubes

RHAC composition by weight RHA:OPC	Wet compressive strength (N/mm^2)	
	7 days	28 days
OPC alone	21	34
1:1	30	44
2:1	15	23

These results, together with other test results above, indicate that good quality RHAC can be made from ash burnt in a heap, but it may be risky to enter into commercial production on the basis of these observations, as only a small disturbance of the burning heap may well affect the temperature profiles, and an unfavourable temperature may occur, with resultant deterioration in quality of the RHAC. It is important that materials produced should have consistent quality, so a more controlled system of burning is preferable.

Malawi

Table 6 shows some of the laboratory test results on ash from several types of incinerator, in admixture with lime, made into mortar.

Table 6. Compressive strength of Malawi mortar cubes

Type of incinerator	Composition by weight RHA:lime	Wet compressive strength (N/mm^2)	
		7 days	28 days
Wire basket	2:1	3.4	–
Brick, adjustable	2:1	2.4	3.7
Brick, fixed vent	2:1	2.3	3.5

There is little difference between the three sources of RHA, though that produced in the wire basket, which was identified above as more pozzolanic than the others, is slightly stronger. Strength of RHAC made from RHA and lime is clearly much lower than that made from RHA and OPC (see Tables 5, 7 and 10).

Guyana

Strengths of mortars made from RHA and OPC are given in Table 7.

The 28-day compressive strength of mortars made with OPC alone is low compared with those in Suriname. Dilution of the OPC with RHA has little effect on strength, even up to twice the weight of OPC. Slower development of strength appears as a characteristic of dilution with RHA.

Table 7. Compressive strength of Guyana mortar cubes

RHAC composition by weight RHA:OPC	Wet compressive strength (N/mm^2)	
	7 days	28 days
OPC alone	14	17
1:1	12	16
2:1	11	14
3:1	7	–

India

Strength of mortars made from RHA from several sources and lime are given in Table 8.

Table 8. Compressive strength of Indian mortar cubes

Method of burning	RHAC composition RHA:lime by weight	Wet compressive strength (N/mm^2)	
		7 days	28 days
Brick/wire	2:1	2.1	4.6
TiB	2:1	8.0	10.5
Pile	2:1	3.4	7.2

The first sample, from the well-controlled commercial production plant to the CRI specification, gave rise to mortars of strength adequate for the requirements of much of the low-cost housing being put up in developing countries, and can be considered the normal standard which could be expected for RHA and lime mixes.

The second sample had been burnt in the Tube-in-Basket, at no more than 750°C, and shows what might be obtained under ideal conditions, though it might be difficult to scale up to commercial production whilst maintaining this quality.

The third sample was made from ash burnt in a pile dumped by truck, not laid down and burnt evenly like the ash in Suriname. Burning is uncontrolled and the product likely to be variable in quality.

Nepal

Cubes were made from two very different types of RHAC in admixture with lime. Strengths are shown in Table 9.

The first sample was made from ash gathered from the small rice roasting plants. Being burnt in small quantities, the husk is not likely to reach an excessively high temperature, the consequence being an amorphous ash (as identified in Table 3), and the high cube strengths which were

Table 9. Compressive strength of Nepalese mortar cubes

Method of burning	RHAC composition RHA:lime by weight	Wet compressive strength (N/mm²)	
		7 days	28 days
Rice roasting units	2:1	6.4	10.6
Massive incinerator	2:1	0.1	0.4

found. This is very small-scale production, but supervision of the subsequent storage and milling of ash was by people thoroughly conversant with the technology.

The other sample is by contrast that made with RHA from the oversized incinerator, with almost no ventilation, which would have attained exceedingly high temperatures, and was found to contain a significant amount of tridymite. The very low strength of the mortar substantiates the comment above that this RHA would be unsuitable for cement production. This method of incineration should not be used. It serves as an example of problems that can arise with scaling up production from otherwise successful units.

Thailand

Mortar cubes were made from OPC and the RHA from AIT and Khon Kaen University; strengths are shown in Table 10.

Table 10. Compressive strength of Thai mortar cubes

RHAC composition by weight RHA:OPC			Wet compressive strength (N/mm²) 28 days
AIT		OPC alone	29
		1:3	28
	approx.	1:2	28
		1:1	18
Khon Kaen		OPC alone	32
		1:4	45
		2:3	41
		3:2	28
		4:1	18

The AIT research shows that RHA can be added in amounts up to 1:2 (ash:OPC), while maintaining strength. Larger amounts of RHA cause reduction in strength. The Khon Kaen work indicates that small additions of RHA increased strength, and ash up to equal proportions with OPC gave no less strength than OPC by itself.

Setting time

The setting time of RHAC is important for the practical purposes of using the material. In Malawi the measurement of initial and final setting times of RHAC made from ash and lime were determined[12] by the standard Vicat method,[8] and showed that RHAC can conform to specifications for similar materials. The initial set was detected at 4 hours, and the final set at 7 hours. This property should be checked in any RHAC project.

Durability

Brickwork laid in RHAC mortar has been exposed to the weather for several years in Malawi and in India, and has been shown to perform well. External renderings on walls in the same countries have demonstrated satisfactory performance. Concrete blocks made with RHAC have been exposed to the weather in Guyana and after a year showed no noticeable deterioration.

Cost

A comparison of costs in India, Nepal and Pakistan, when RHAC was being produced at several centres during 1982, showed that at that time RHAC was significantly cheaper than OPC. On average, the selling price of RHAC was 45 per cent of the price of OPC.[5]

Conclusions

RHAC has been widely researched and although there are differences in the quality of the product due to the variation in the methods of production, it is clear that with proper design of the production plant and method, a good quality product can be made.

Rice straw does not form the basis of a good pozzolan.

RHA taken from a burning pile at the rice mill can be satisfactory for use in making RHAC, if the husk is burning evenly and continuously near the surface of the cone formed as husk is ejected from the mill. However, as this condition may be difficult to guarantee, controlled burning in a special incinerator is generally advisable.

Laboratory analysis shows that high temperatures during the incineration of rice husk result both in the formation of crystalline phases of silica in RHA and in low-strength RHAC. Temperatures below 750°C are preferred, as being conducive to formation of the more reactive amorphous silica ash, and higher strength RHAC. Pink or lilac-coloured ash is likely to be very crystalline, and of little use in making RHAC.

Incinerators for burning rice husks should allow moderate access for combustion air to all parts of the burning husk. However they should be protected from excessive draughts, which may otherwise lead to fast burning and high temperatures. Incinerators should allow heat of combustion to escape so that the temperature does not rise too high.

A ball mill is eminently suitable for grinding RHA to powder. Pestle and mortar and maize mill are not so effective as a ball mill.

RHAC can be made with RHA and hydrated lime. In the proportions 2:1 respectively, wet compressive strengths of mortar of the order of $4N/mm^2$ may be anticipated.

RHAC can be made with RHA and OPC. RHA up to an equal weight of the OPC may give mortar with strengths at least as high as if OPC were used without dilution with RHA. Wet compressive strengths of such mortars may be from 20 to $40N/mm^2$.

Time required for RHAC to reach its first set can be sufficiently long to permit easy use of the material, while final set can be quick enough to allow construction to proceed at an acceptable rate.

Production of RHAC is most likely to be feasible in countries where OPC is in short supply or expensive. Countries which do not possess raw materials for production of lime or OPC, whilst they do have large outputs of rice, stand to benefit most from RHAC production. The price of RHAC can be lower than that of OPC.

Durability of RHAC in mortars, in renderings, and when made into concrete blocks, has been shown to be good.

Tests should be carried out with local materials and under local conditions before entering upon large-scale development of rice husk ash cement production.

11. Research on development of alternative cements based on lime pozzolanas in Uganda for use in rural housing

W BALU-TABAARO*

In developing countries and in Uganda in particular about 70 per cent of the population live in inadequate housing. According to a study carried out by Uganda's Ministry of Housing Urban development (National Housing Programme 1987 to 2000), in 1986 the rate of annual construction required from 1987 to 1990 was estimated at 253 750 units. Rural areas will need an annual construction rate of 38 000 units during the period 1987–90.

---

*Department of Geological Survey and Mines, Entebbe, Uganda.

In Uganda, concrete blocks and mortar are manufactured using Portland cement. These materials are in short supply and sold at prohibitive prices. This situation calls for a low-cost replacement for Portland cement which could reduce the cost of housing.

Local raw materials for the production of lime-pozzolana cements in Uganda

Limestone occurs in Uganda in large quantities. Large quantities of pozzolanas are also available and these can be categorized into two groups:

- natural pozzolanas which include volcanic ashes, and tuffs, pumice, scoria, obsidian and others
- artificial pozzolanas which include fly ash, burnt clays, waste lime sludge, and agricultural wastes.

Pozzolanas of volcanic ashes, scoria and tuffs occur extensively in the Kisoro areas. In the same area, there is wolfram mining and the tailings of these mines contain good quality volcanic ash and scoria, though their chemical properties have yet to be assessed. Sediments of ashy materials occur in Bunyaruguru–Kazinga channel and Fort Portal areas while diatomaceous earth occurs in the Pakwach areas.

Pozzolana of plant origin

Pozzolanic materials of plant origin include coffee husk ash, bagasse ash, groundnut shells, legumes, agricultural wastes. Rice, coffee, sugar, groundnuts (peanuts) and legumes are extensively grown in Uganda and their husks could be used to produce ash for use with lime.

A UNIDO estimate puts the annual production of bagasse at 850 000 tonnes, coffee husks at 136 000 tonnes and rice husks at 10 000 tonnes.

Soft-fired clay from reject bricks

Uganda has abundant clays which are burnt for the production of bricks and tiles. The rate of reject for these products is quite high and these reject bricks and tiles can be crushed and ground for use in lime pozzolana cements. One factory near Kampala, Uganda Clays, produces about 30 000 tonnes of bricks per year with an appreciable amount of rejects, estimated at 2400 tonnes per annum.

Agro-industrial ashes

Agro-industrial ashes are available from a number of industries in Uganda. These include bagasse as a source of heat. Other industries include the beer industries, tobacco factories, coffee processing plants and meat canning industries. Most of the factories are fired by wood, agricultural wastes, bagasse and oil. Large amounts of ash are produced as by-products and sometimes these cause disposal problems. The use of such ashes could provide cheap cement and their disposal would help in preventing environmental degradation. Qualities and quantities of these sources are yet to be assessed.

The use of volcanic ash from the Virunga Mountains in the development and manufacture of lime pozzolana cements

Because of the high cost of Portland cement, it was found necessary to carry out detailed research on the development and utilization of the abundant volcanic ashes in south-west Uganda in order to create a cheap alternative cement. With financial assistance from IDRC of Canada, and expertise from the Department of Civil Engineering of the University of Toronto, led by Prof. R.H. Mills, detailed work on volcanic ashes is being carried out. The main objective of this project is to increase production of lime pozzolana cements so that there is increased access to it by the low-income and rural-based populace in Uganda.

Based on the preliminary geological literature survey, various samples were collected from the Muhavura Mountains and lime samples from Kasese and Busanza. The volcanic ashes were analysed both in Toronto and in Uganda and their chemical analysis is shown below. The lime samples were analysed locally and their results follow.

Muhokya lime

The complete assaying of Muhokya lime is being carried out but results of the rapid sugar test for determination of available lime revealed a low figure of 22 per cent for Muhokya lime. This lime contains impurities such as charcoal sand and organic matter. The specific surface area using a Blaine air permeability apparatus was found to be $1640 cm^2/g$. Appropriate

Table 1. Tororo lime

	Percentage
SiO_2	1.00
AlO_2	1.30
Fe_2O_3	3.20
CaO	63.69
MgO	2.80
Loss on ignition	21.41
Others (insoluble)	4.41

Table 2. Kaku lime

	Unburnt	Burnt samples
CaO	51.77	64.19
Fe_2O_3	7.08	2.65
MgO	2.03	2.58
CO_2	41.22	10.75
H_2O	2.24	16.59
Insoluble acids	1.13	3.62
	100.47	100.41

All the iron has been calculated as Fe_2O_3. FeO was not determined separately.

Table 3. Silica and alumina content in volcanic ash

Sample	SiO_2	Al_2O_3	Total
1	38.3	12.7	51.0
2	41.2	12.1	53.0
3	47.0	14.8	61.8
4	43.8	11.8	55.6
5	46.4	14.6	61.0
6	43.7	13.2	56.9
7	45.2	13.7	58.9
9	42.8	10.0	52.8
9A	45.5	13.2	58.7
10	42.3	10.2	52.5
12	39.7	9.2	48.9

Minimum $SiO_2 + Al_2O_3$ required is not less than 60 per cent. Thus, the most suitable volcanic ash is in samples 3 and 5 (Nyagishenyi and Hakirembe respectively).

methods are being devised to carry out on-the-spot tests to ascertain the quality of the lime before use in pozzolan cement.

Natural particle size of volcanic ash

Volcanic ash in the Virunga Mountains occurs in small particles not exceeding 30mm except when it is in an aggregated form.

Sieve analysis was carried out on the bulk samples before grinding. The particle size distribution of at least four samples was determined. The results obtained are summarized in Table 4.

Crushing and grinding tests

Three types of crushing or grinding equipment were tried out for testing the volcanic ash samples as shown in Table 4.

Table 4. Screen analysis of volcanic ash in its natural state

Sample and origin	Distribution Function	Sieve size in μ					80% passing sieve size (μ)	50% passing sieve size (μ)
		12700	6350	3180	1000			
PB1	% retained	25.3	33.3	27.0	9.5		14000	76000
	cumm % undersize	74.7	41.4	14.4	0.9			
PB2	% retained	25.7	22.3	26.8	19.7		14000	6000
	cumm % undersize	74.3	51.9	25.1	5.4			
PB3	% retained	0.6	0.5	31.1	49.9		4700	2300
	cumm % undersize	99.4	92.9	61.8	11.9			
PB4	% retained	4.8	30.9	37.7	23.3		9000	4800
	cumm % undersize	95.2	64.3	26.6	3.3			
MUKO	% retained	0.2	9.7	43.9	40.9		5200	3300
	cumm % undersize	99.8	90.1	46.2	5.3			

Crushing and grinding tests
Three types of crushing or grinding equipment were tried out for testing the volcanic ash samples as shown in Table 4.

Table 5. Crushing and grinding test (specification of equipment used)

	Ball mill 8⅔"× 8"	Roller mill 12½× 6⅙	Jaw crusher 7½× 5½
Power	1 PH = 0.75kw	2 PH = 1.5kw	3 PH = 2.25kw
Voltage	400–440V	400–440V	415V
Current	1.85A	3.7A	4.76A
Speed	–	1139 RPM	720 RPM
Phase	3		
Volume	7190cm³	6640cm³	2925cm³
Mill wt	19.4kg	14.5kg	3.9kg

Table 6. Physical properties of volcanic ash

Sample	Specific gravity	Moisture content	Work index kWh/tonne
PB1	2.22	2.53	10.16
PB2	2.30	0.39	12.23
PB3	2.43	0.998	11.72
PB4	2.40	0.884	8.77
MUKO	2.38	0.957	9.59

Physical properties of volcanic ash

Specific gravity, moisture content and work indices of volcanic ash samples were determined. The results are shown in Table 6 and a brief description of the methods is given below.

Specific gravity

The specific gravity of volcanic ash was determined using the elementary water displacement method. The results obtained agreed satisfactorily with those of similar materials: pumice (1.96), shale (2.88), diatomite (2.65), limestone for cement (2.69), tile (2.59).

Moisture content

Samples were weighed before and after drying in an oven for five hours. The difference in weight gave the moisture content.

Work index

By knowing the work indices of some materials, various volcanic ash samples were ground under similar conditions to those of known materials. From the test results, work indices of volcanic ash were determined.

Test on lime-pozzolan trial mixes

Using the different samples of volcanic ash and lime, trial mixes were carried out. Volcanic ash was crushed in a roller mill, then later ground in a ball mill for four hours and finally interground with lime for one hour. The resultant mixes were tested for fineness, shrinkage, consistency and setting times. Replacement of some pozzolanas with ordinary Portland cement was also done. Sand mortars were prepared and tested for strength development with time after curing for 7 days and 28 days. Results of tests carried out so far are illustrated in Table 7.

Stabilization tests

Sample 3 pozzolan cement was used to stabilize the lateritic soils and the mixture used to make building blocks. The compressive strengths of the blocks were tested weekly; the results are shown in Table 8.

Pozzolan-cement mortars

Pozzolan-cement mortars are being tried out and tests are still continuing. The results so far obtained are given below.

Mortar cubes for Nyagishenyi pozzolan were produced and tested for compressive strength. Muhokya and Italian lime were used in lime-pozzolan cement production.

Cement type	Mixture Ce : Sa : Wa			Compressive strengths (MPa)
Nyagishenyi (Muhokya)	1	3	0.6	0.3
Nyagishenyi (Italian)	1	3	0.6	0.5

Wall assemblies

From the data so far obtained, wall assemblies were designed and constructed to assess the field performance of the lime-pozzolan cement mortars and soil stabilized blocks.

Wall assemblies were put up using soil blocks stabilized with lime-pozzolan cement. The details of the assemblies are given below:

Materials used: Soil stabilized blocks at 8 per cent stabilization.
Cement type: Lime: pozzolan (Muhokya:Nyagishenyi, 1:3)
Block dimension: 29 × 8.3 × 13.8cm
Mortar mixture: Cement:Sand:Water (1:3:0.6)
Wall dimensions: Height: 63cm
   Length: 122.5cm
   Thickness: 22.5cm.

A second wall with similar dimensions was put up utilizing Nyagishenyi pozzolan and Italian lime. The pozzolan cements were used for binding and plastering of the walls erected outside, exposed to various weather conditions. There was little degeneration observed of either type over a 10-week period.

Table 7. Pozzolan cement tests

Cem. type	Pozz. sample	Lime	Fineness cm²/g Blaine	Standard consistency %	Initial set (min)	Linear shrinkage	Compressive strengths MPa (7 days) 8 × 4 s. cylinder
PozzIII	3	Muhokya	5000	26	210	1.75	2.6
PozzIV	3	"	4900	22	220	1.85	
PozzIII	3	"	4900				2.57
PozzIII*	3	"	4900	24	165	1.92	6.03
PozzIV*	3	"	4900	20	140	1.96	
PozzIII	3	Italian	4618	24	240	3.14	
PozzIV	3	"	4120	26	260	3.14	
PozzIII*	3	"	4618	25	130	2.1	
PozzIV*	3	"	4120	26	160	0.74	
PozzIII	5	Muhokya	4995	22	360	1.85	
PozzIV	5	"	5000	24	420	1.92	
PozzIII*	5	"	4995	24	130	1.6	7.56
PozzIV*	5	"		26	170	1.8	
PozzIV*	5	"	3360				8.45

(paste water cured for 14 days)

Note: PozzIII refers to Pozzolan:Lime ratio 3:1 and the same applies to PozzIV.

* These pozzolan cements have a 10 per cent replacement with Portland cement i.e. OPC:Lime pozzolan cement, 1:9.

Table 8. Stabilization tests

Quality of stabilization	LPC:Laterite soil
	1:4
Blaine fineness of LPC used (cm²/g)	4000
Moisture content at the time of casting	18%

Brick mark	Wet weight (g)	Density (kg/m³)	Dry weight (g)	Density (kg/m³)	Age of brick (days)	Compressive strengths (MPa)
A	8500	2373	7750	2164	7	4.52
B	8300	2318	7300	2038	14	4.4
C	8000	2234	7000	1955	21	4.2
D	5200	1857	4900	1750	28	3.0

Table 9.

Quality of stabilization: LPC III:Laterite soil 1:8
Blaine fineness of LPC used: 3800
Moisture content at the time of casting: 16%

Brick mark	Wet weight (g)	Density (kg/m³)	Dry weight (g)	Density (kg/m³)	Age of brick (days)	Compressive strengths (MPa)
12	7800	2178	7200	2010	7	3.4
13	7950	2220	7350	2053		3.5
14	7850	2192	7100	1983	14	3.2
15	7500	2094	6800	1899		2.4
16	8550	2388	7700	2150	21	3.9
17	7550	2108	6800	1899		2.5
18	7800	2178	7000	1954	28	3.25
19	8200	2290	7350	2053		3.75
20	7700	2150	–	–	30	–
21	7800	2178	–	–		–

Table 10.

Quality of stabilization   LPC III:Laterite soil
Blaine fineness of LPC used   1:6
    3800cm²/g
Moisture content at the time of casting   17.4%

Brick mark	Wet weight (g)	Density (kg/m³)	Dry weight (g)	Density (kg/m³)	Age of brick (days)	Compressive strengths (MPa)
1	8250	2304	7450	2080	7	3.5
2	8300	2318	7450	2080		3.5
3	8150	2276	7150	1996	14	3.0
4	8050	2248	7050	1969		3.0
5	8350	2332	7250	2025	21	3.7
6	7800	2178	6850	1913		3.0
7	7700	2150	6900	1926	28	2.5
8	8100	2262	7100	1983		3.4
9	7600	2122	6600	1843	30	2.6

Comments on the tables

There is a high positive correlation between the density of a brick and its compressive strength, i.e. the higher the density the higher the compressive strength.

The quality of stabilization does not have a significant impact on the compressive strength of the brick but is expected to have a major influence on the permeability of the bricks. Permeability was not assessed due to lack of the necessary equipment, especially the permeameter.

Pozzolan cement with aggregate

Utilization of lime-pozzolan cements in concrete yielded poor results. Mixtures in the ratio of cement:sand:aggregate, 1:3:4 were used with a water-cement ratio of 0.6. Thus a scheme to replace some of the pozzolan cement with ordinary Portland cement was embarked on. This is illustrated below with the results of some tests so far carried out.

Table 11. Concrete tests

Nyagishenyi–Muhokya 1:3
Cement:Sand:Aggregate 1:3:4
W/C = 0.6

OPC	Pozz LPC	Compressive strengths	
		3 days	7 days
100	0	–	–
90	10	–	–
80	20	5	5.5
70	30	6.1	–
60	40	–	–
50	50	2.8	–

Utilization of lime-pozzolan cement in fibre-concrete tiles

Fibre-concrete tiles have been produced with Portland cement replaced by lime-pozzolan cement at Namuwongo Co-operative Society. The mix ratios used were:

OPC	:	Lime-pozzolan cement
90		10
80		20
70		30
60		40

In the manufacture of tiles, the mix of cement:sand:sisal was 1:3:0.04. The tiles were water-cured for 7 days after air drying for 24 hours. Further testing is to be carried out on the tiles to assess their performance. In this investigation Nyagishenyi pozzolan was used but trials on other pozzolanic materials will be made.

Conclusions

Tests on lime-pozzolan cements showed no significant difference between LPC (Muhokya) and LPC (Italian) with regard to consistency and setting time.

Blending of some Portland cement reduces the percentage shrinkage i.e. linear shrinkage of Pozzolan III and Pozzolan IV at 3.14 per cent which was outside the BS and ASTM limit of 3 per cent. This was rectified to 2.1 and 0.74 per cent respectively by a 10 per cent replacement of LPC by Portland cement.

It is also evident that OPC activates LPC with regard to the setting time and thus enhances the rate of strength development.

There is negligible change in strength development with increase in fineness above 4000cm^2/g Blaine.

The LPC tested did not comply with the BS test for soundness. Only those LPCs blended with OPC met the soundness test.

The pozzolan cements exhibited low strengths because of the inadequacy of the lime. Muhokya lime contained impurities such as charcoal, sand and organic matter. The available lime index was only 22 per cent. This suggests that the cement that was being produced had a great proportion of unreacted pozzolan.

Both the Muhokya and Italian lime used had suffered some carbonization, limiting the reactive centres available for interaction with the pozzolan and then later for the hydration reaction.

The fineness of the lime was low (c.f. ASTM recommendation for limes to be used in pozzolan cements, viz. 10000cm^2/g Blaine, lime used in the tests: 1640cm^2/g Blaine). A higher fineness could be achieved during intergrinding but only if it is ascertained first that the quality of the lime is high enough.

Assessment of the results from the current work shows that there is great potential for the use of lime-pozzolan or lime-pozzolan blended with Portland cement or even Portland-pozzolan cement.

Improvement on the quality of lime used is a major prerequisite for better performance of pozzolan cements. Better kilns will have to be built to produce lime of CaO content of at least 65 per cent.

Addition of other testing facilities is vital to allow for more in-depth testing of these cements to come out with specifications and standards for Uganda pozzolan cements. These include a permeameter, more test cubes, cylinders and pH meters.

A fibre-concrete tile making and transverse strength testing machines are urgently required to facilitate further research into incorporation of pozzolan cements in the tile industry. This would reduce the cost of the tiles making them affordable by low-income earners of Uganda. These tiles could be used in the roofing of model houses to be erected in field assessments of performance of pozzolan cement as a material of construction.

Recommendations

The optimum ratio to be adopted for mixing pozzolana and lime is 3:1. A minimum fineness of $3600 cm^2/g$ should be adopted for field trials.

Appendix

Description of samples

P1 – Hakadege, Kisoro – Bunagana Rd: Ash consists of yellowish brown layered formation.
P2 – Karambi – 4km on Kisoro – Kyanika Rd: Darker and uniform grained.
P3 – Gatarara – 1km from Kyanika on Kyanika – Kisosro Rd: Ash and tuffs are finer and lighter coloured.
P4 – Nyakabande – Kilembe: Greyish black ash with isolated boulders.
P5 – Nyakbande – Catholic: Same as P4.
P6 – Kamageza: Finer grain weathered to yellowish grey at top.
P7 – Kaseregenyi: Large boulders brownish grey – Pumice lava.
P8 – Katozo – Kalengire: Yellow brown lava.
P9 – Katozo III (lateritic conglomerate): with lava.
P10 – Muko – Kasenyi junction: Dark fine ash.
P11 – Muko Parish: Same as P10.
P12 – Muko III: same as P10.
P13 – Muko IV – Muko market: same as P12 but have yellow cap at top.
P15 – Kaseregenyi – Murram pit: contains gravel and pebbles.
P16 – Hakadege II: Dark grey to black gravel with sand, light grey.
P17 – Hakadege – Murram pit: purplish with gravel and sand with flakes of mica.
P18 – Dark grey to black uniformly grained ash.

PB1 – Bunagana Rd.
PB2 – Kyanika Rd.
PB3 – Muko Catholic Mission.
PB4 – Kikomba.

N.B. All the samples above are from Kisoro district in S. Western Uganda.

12. Costa Rica: Small wood-fired kilns
JOSE PACHECO*

Most of the lime produced in rural areas of underdeveloped countries is made in batch kilns. Many of them have their walls underneath the earth in order to insulate the heat, and because this is the cheapest way to build it. The kiln is a simple cylinder made of rock or bricks with a hole in the bottom fired by wood. The same hole is used to burn the lime or to discharge the kiln. After each burn, a man has to go inside the kiln to discharge it. A kind of dome has to be made to support the new charge. Working conditions are very bad because of the high temperatures.

Lime kilns in Central America and Costa Rica

In Central America, there are approximately 800 traditional kilns. They produce 90 per cent of all the lime in these countries. In Costa Rica, a tropical country with a population of three million inhabitants, and an economy based on agriculture, lime continues to be a small rural industry. We have about 45 of these traditional kilns.

Table 1. Chemical composition of carbonate in different areas

Site	$CaCO_3$	SiO_2	$MgCO_3$	Al_2O_3	Fe_2O_3
Abangares	98.0	0.35	0.36	0.26	–
Nicoya	98.0	0.35	0.36	0.26	–
Rio Azul	95.6	2.60	0.84	0.72	0.65
Patarrá	86.3	8.50	0.73	1.72	1.71

Reference: Cementos del Pacifico C.R.

Costa Rica consumes about 16 725 tonnes of lime per year. There are four different regions that have good conditions for producing lime. Lime production is concentrated in two of them: Patarrá and Nicoya. The other two are used only to mill the carbonate. Information on the size of these deposits is lacking because small producers do not have the resources to do systematic studies.

Most of the lime-producers combine this activity with coffee growing. An average of 25 per cent of the firewood is obtained from their own coffee plantations. The rest is obtained from sawmills that work with wood from natural forest. However, in four or five years, these sawmills are going to be forced to close, because Costa Rica has strong environmental legislation

---

*Instituto Tecnologico de Costa Rica, Cartago, Costa Rica.

with the purpose of protecting the natural forest, which comprises 30 per cent of our territory. In addition to this legislation, funds are invested in reforestation for industrial purposes. From these plantations, enough thinning wood could be obtained for the lime industry. However, the main problem in using this source of wood is that many people continue to cook with wood instead of electricity due to its low cost. Firewood plantations are not profitable because the price of the wood is still too low.

Table 2. Cost of industrial fuels

	Energy unit cost ($)	Cost/energy usable ($/Gj)
Diesel	0.28/lt	8.82
Bunker	0.15/lt	4.04
Firewood	4.66/m^3	0.77
Propane	0.45/kg	9.49
Liquid gas	0.16/lt	6.72
Electricity	0.26/kWh	7.40

Reference: Proyecto Hornos de Cal.

Production groups

Small producers in Patarrá make an average of 12 burns a year in each kiln during the dry season from December to March. They have a very traditional approach to limemaking. The whole process is done manually, including hydration.

The other area, Nicoya, burns lime during the whole year. They use the same kind of kiln as the Patarrá producers, but they complement this process with extraction-milling, classification and mechanical packing. Only the hydration process is manual. They are increasing their markets and are concerned about the quality of their product. They also realize that if they do the hydration process mechanically, they will be able to establish quality standards.

Table 3. Productive organization

Site	Location	Producers	Use
Nicoya	Rural	3	Industrial
Patarrá	Semi-rural	33	Agricultural

Ref: Proyecto Hornos de Cal, 1988.

Technology of batch kilns

The technology of batch kilns based on local materials and local knowledge is at risk because the process has not been changed in hundreds of years. The main advantages of this technology are:

- production for local markets
- low transportation cost
- low investment ($5714)
- good calcination quality
- pre-existing infrastructure
- local materials available for construction.

The main problems are:

- poor working conditions
- low productivity
- low energy efficiency
- the mixture of lime with ashes.

The Instituto Tecnológico of Costa Rica (ITCR) and Appropriate Technology International have developed a new model considering all of these variables. This modified kiln developed by the ITCR has approximately the same geometric proportions as the traditional one (diameter of 2.5 metres and 5 metres high). However, we have developed a permanent structure which supports the charge and allows the heat to run freely through it. The main differences of this kiln in relation to the traditional one are:

- its cost is only 20 per cent higher than the traditional one ($7142) for the same volume of lime
- good working conditions
- easier and faster charging and discharging system
- no contamination between lime and ash
- 10 per cent less wood consumption
- 40 per cent higher productivity.

A traditional kiln can be transformed by introducing these changes if it has a slope of 45 per cent in front of it or into one of its sides. The following

Table 4. Comparison between traditional and improved batch lime kilns

	Traditional	ITCR/ATI
Volume of the kiln (m^3)	21.46	24.87
Weight of firewood (kg)	23 674	24 189
CaO (kg)	17 138	20 365
No. of burns per month	4	5
Production of CaO per month (kg)	68 552	101 825
Cost of the kiln (US$)	5 714	7 142
Stone/preparation (hrs/person)	266	333
Charging process (hrs/person)	23	22
Kiln charging (hrs/person)	33	34
Kiln discharging (hrs/person)	32	15
Total (hrs/person)	357	407

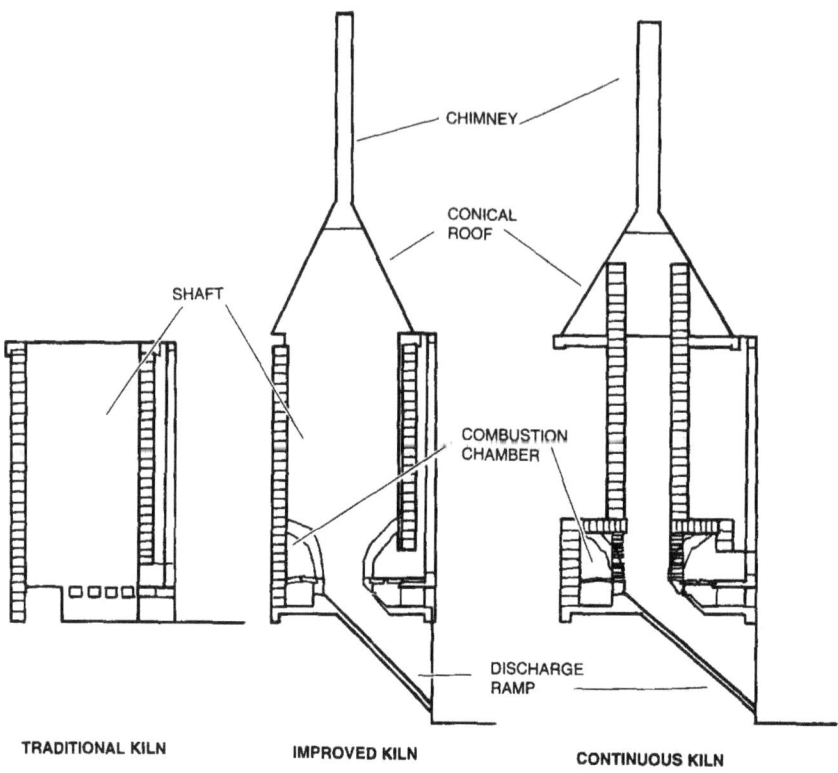

Figure 1: *The traditional kiln has been adapted to facilitate unloading*

characteristics of the improved kiln allow it to be used as a continuous kiln:

- heat is homogeneous
- the discharging process uses gravity, which can be controlled
- the only part of the kiln (ITCR/ATI) that has to be changed in order to transform it into a continuous kiln are the proportions of the walls.

The continuous kiln and the lime market

During the last 30 years, there have been five attempts to operate continuous kilns in Costa Rica and all of these were ruined by the size of the market. Half the extracted limestone in Central America is milled for agricultural purposes and the other 50 per cent is used for lime, but half of this lime is also used in agriculture.

It is not logical to promote the use of lime in agriculture if it is possible to use calcium carbonate with lower production costs. For that reason, the only way to increase the lime market in Costa Rica is to produce lime for industrial purposes. Our goal in the development of this kiln was to use our infrastructure in a more rational way and to create the conditions to transform our industry. This has to be complemented with the rest of the processing stage. We are now interested in improving the use of the most adequate processing systems that exist for small-scale production, specially hydration systems.

Some of the medium-size producers have incorporated milling and classification systems in their industry but they continue hydrating the lime without machinery. Small producers do not use any machinery at all.

This situation prevents us from promoting standards for different industrial uses of lime for these small industries. In Costa Rica, the price of lime used for industrial purposes is almost five times the price paid for agricultural lime.

Our market studies show that it is possible to sell this product in industrial markets and the production of this small kiln can be standardized if it is used appropriately.

ITCR is going to participate next year in a demonstration geared toward the insertion of these technological changes with small producers in the Central American Region. If this goal is reached, the new problem could be the need for more productivity. In this case, kilns could be modified by changing them from batch to continuous.

13. The lime industry in Zimbabwe

PHILEMON NHACHI*

The limestone reserves of Zimbabwe, although not completely surveyed, are sufficient for present demand and for many years to come. The problem connected with Zimbabwean limestone deposits is not one of quantity but of maldistribution and quality; the location of suitable limestone does not coincide with the density of the population. Remoteness and distance from the railway are serious obstacles in exploiting many promising deposits, especially of calcrete. In Zimbabwe, limestone occurs in almost all the major geological units. Approximately 90 per cent of the deposits are in the early Precambrian Bulawayan System of the Greenstone Series. These deposits are formed in lenticular masses lying in close conformity with

---

*G&W Industrial Minerals, Harare.

banded ironstones and are normally very impure, with unacceptable amounts of silica, iron and other impurities.

The mid-Precambrian sedimentary basins of the Lomagundi, Deweras, Piriwiri and the Umkondo contain the other 10 per cent of known deposits. The late-Sijarira sediments and post-Karoo alkali ring complexes contain insignificant deposits.

A total of 119 limestone deposits exist throughout the country. The producing deposits are as follows:

Contact in Bindura, Copthal in Colleen Bawn, Dodge in Shamva, Early Worm in Concession, Falcon Limes South of Kadoma, Jack near Bulawayo, L.M. Gwati near Chishanya, Mafungabusi in Tshombe, Paignton in Colleen Bawn, Ripple Creek in Kwe Kwe and Sternblick in Harare.

Products and uses

Production of cement in Zimbabwe is undertaken by two companies. Circle Cement Ltd operates the Sternblick Quarry just east of Harare and United Portland Cement Ltd operates the Colleen Bawn and Bulawayo factories, using limestone from the Paignton and Copthal Quarries in Colleen Bawn.

Limestone from various producers – notably Early Worm, Bloomfield, Dodge, Jack, Tuli, Pioneer, Redcliff and various other deposits – is used in many applications such as the ferrochrome and steel industries as a flux, agriculture, building, fillers in rubber, paints and many other industries.

Most of the limestones in Zimbabwe have high iron and silica contents and a correspondingly low lime content. As a result, very little burnt limestone (or lime) is produced to meet specialized metallurgical process industries. Such lime is imported as great expense.

As various sectors of the economy expand, more and more limestone products will be required. Great process innovations will be required to make some of the limestones usable.

Lime production in Zimbabwe

The typical situation of a lime plant in Zimbabwe is depicted in the example of Early Worm Mine lime works which is an open cast mine situated 16km north of Concession. It was first pegged in 1957. The prospects of the mine have oscillated during its history. The mine had for the most part supplied calcitic limestone for the agricultural sector with a very small percentage being sold for industrial uses, despite the high quality of the material available (98 per cent $CaCO_3$).

After the successful acquisition of the Early Worm Mine by G&W Industrial (Pvt) Ltd, a wholly owned subsidiary of a parastatal, the Industrial

Development Corporation of Zimbabwe, the mine was developed into a multi-product line lime plant. This was achieved by calcining limestone as well as grinding it into various lime-based products. The products currently produced by the plant are: agricultural lime, stock feed lime, super fine lime, chemical lime, slaked lime, building lime, chicken grit and ¾-inch (20mm) stone.

Limestone is drilled, blasted and excavated from the surrounding area and converted into the oxide through the following processes:

Drilling
Ordinary compressors and drilling equipment are used for this purpose.

Blasting
This is effected in the normal manner whereby high explosives are employed.

Excavation
This is done by bulldozers and manual labour into tractor-dumpers by a frontend-loader.

Calcining
The selected crushed 4–6 inch (100–150mm) limestone material is hoisted to the kiln top using conveyor belts and fed at the top of vertical shaft kiln. Coal or coke is utilized for burning at a limestone/coke ratio of 5:1. Calcination is achieved by burning the mixture at 1000°C.

Slaking
Water is sprayed over the discharged calcined crushed lime to hydrate the burnt lime.

Grading
Fine powder of lime is obtained by manual sieving and vibratory screening as well as by air separation.

Bagging
The fine lime powder is manually or mechanically filled in 25kg 3-ply paper bags, weighed and stored or distributed to consumers.

Production capacity
The three vertical shaft kilns installed at Early Worm Mine were designed and fabricated by G&W personnel. They were fabricated out of disused

asbestos fibre dryers purchased as scrap from an abandoned asbestos mine. The three kilns are 18m high, one with a diameter of 2.3m and the other two 1.7m diameter. The shells of the three kilns are made of ¾ inch (20mm) cast-iron rings. The kilns are lined with 9-inch (230mm) refractory bricks separated by an inch of raw vermiculite from the shell. The plant has a design capacity of +/–40 000 tonnes per annum of hydrated lime whose marketable value at about Z$130 per tonne realizes an annual turnover of over Z$5 million.

Uses

All burnt lime from the Early Worm Lime Kilns is used to produce hydrated lime which in turn is used for various purposes, mainly as pH modifier in gold and copper mines. The main applications are as follows:

Soil stabilization for roads

The standard method of making a road is to replace the soil underneath to a predetermined depth with hard (stone) aggregate on which a special surface is finally laid. It has been found that by admixture with relatively small amounts of hydrated lime (determined by laboratory tests) the original soil in the base can be stabilized saving the cost of much of the depth of aggregate which was formerly required.

Building and construction

Together with sand, hydrated lime forms the basis for mortar and plaster.

Cement extender

Hydrated lime is mixed with Portland cement PC 15 in the ratio 3:1 hydrated lime to cement.

Lime consumption in Zimbabwe

Table 1 shows the demand pattern for lime in Zimbabwe.

The main industries using lime in Zimbabwe are: cement, building and construction, iron and steel, sugar, low carbon ferrochrome, and the mining sector.

The cement and iron and steel industries have their own limestone deposits which are large enough to supply their own demands for some considerable time, as well as their own calcining facilities.

The low-carbon ferrochrome industry requires very high quality lime to use as a fluxing agent in the production of low-carbon ferrochrome. The

Table 1.

Industry	Off take ('000 tonnes)
Cement	1000
Agriculture and animal feed	100
Low carbon ferrochrome	65
Iron and steel	65
Municipalities	12
Sugar industries	12
Mining sector	10
Pulp and paper	8
Building and construction	6
Glass industry	5
Leather industry	1
Total	1287

stipulated specifications of the lime are: calcium oxide content 92 per cent, silica content less than 2 per cent, iron content less than 1 per cent, magnesium less than 1 per cent and alumina less than 1 per cent. The low-carbon ferrochrome industry imports over 60 000 tonnes per annum from South Africa.

The municipalities also use high quality slaked lime in water treatment plants. The magnesium content in this lime should be as low as possible because it reduces the coagulation efficiency of lime. The annual requirement of municipalities for lime is +/-8000 tonnes and they import it from both South Africa and Zambia.

Sugar industries use lime as a coagulant and neutralizer. Their annual requirement is +/-6000 tonnes and this is also imported from South Africa.

The copper and gold mines use +/-6000 tonnes of lime per annum. Other minor consumers are building and construction industries and the agricultural sector.

The annual import of high quality lime is 100 000 tonnes and it costs the nation Z$5 million per annum (at Z$130 per tonne). This gap in the market could be filled by locally produced lime if a suitable deposit could be located.

The consumption of imported lime is mainly by those industries which have relatively stringent quality requirements. These industries are low-carbon ferrochrome, municipalities, sugar, glass and pulp and paper.

Metallurgical industry

The metallurgical industry has specific chemical purity requirements for limestone and lime. The major portion of lime produced in Zimbabwe is by the simple shaft kiln method and heap burning with no quality control procedures being followed. Therefore the lime that is used in the metallurgical industry in Zimbabwe has to be imported.

Iron and steel

Lime is used in the manufacture of pig iron and steel as a fluxing agent and for purifying the metal by forming a molten slag that is removed by utilizing density differences. The calcium (Ca) and magnesium (Mg) form complexes of silicates, phosphates and sulphides in the slag.

Limestone is universally employed either in lump form in the blast furnace for iron making or in a finely subdivided state to produce ore agglomerates, like self-fluxing sinters and pellets. In the finely subdivided state in the iron ore concentrate, the flux charge may be supplemented with quick or dry hydrated lime.

Agriculture

The use of lime and limestone dust in the agricultural industries entails very loose quality restrictions compared to the metallurgical industries.

Limestone dust and lime are used as soil conditioning agents by stabilizing soil acidity. Lime is preferred in this context; acting faster with higher neutralizing value per tonne than limestone. However it is more costly than limestone.

It is also used in fertilizer blending and mixing plants where it is utilized in the form of pulverized limestone. Dolomitic limestone is preferred as a fertilizer filler rather than pure calcium carbonate. As a filler, lime has not found extensive use in Zimbabwe.

Animal feed

Calcium, closely followed by phosphorus, is required in larger quantities than any other mineral in the animal diet. Typically, 35 per cent available calcium is needed as a mineral supplement. Pulverized limestone is preferred to lime because it provides readily available calcium most inexpensively. It is also ideal as it is commonly a carrier for the other minerals and trace elements that constitute the imported mineral supplements.

Of specific use for poultry are grits of limestone origin. The important quality parameter in this case is the size. Typically hens require about 0.44cm size. Grits need to be very carefully screened so as to achieve as little particle gradation as possible.

In Zimbabwe 7100 tonnes/year of pulverized limestone are used as a component in animal feed formulae. The demand will undoubtedly increase with the inevitable growth of the poultry industry.

Glass and ceramics

Limestone or quicklime and dolomite are used interchangeably as raw material sources of CaO and MgO in the glass industry.

In the manufacture of basic refractory bricks, such as used as heat insulators in cement kilns, dolomite is the main raw material for periclase MgO.

The quantities of lime and magnesia entering the batch in the raw and burnt materials are about equal. A good limestone for glass making will contain 30–56 per cent CaO and up to 21.5 per cent MgO.

Sugar

Prior to crystallization into granulated sugar, the liquor is normally purified from the melter. The purification process commonly used involves clarification, filtration, de-colouring and ion exchange.

Clarification is achieved by chemical treatment of the liquor with lime. The process liquor is 'limed' to neutral pH using milk of lime and heated.

Water treatment

Lime is jointly applied with soda ash in water softening. By retention of water with an excess of lime for 24–28 hours, bacteria are killed in addition to removal of the temporary hardness.

Lime is also used in water pH control where coagulation of suspended or non-settling solids from turbid waters is effected through the use of coagulants such as alum and iron salts.

Other industries

Limestone is used as the major raw material in the cement industry. The consumption of limestone in Zimbabwe for cement manufacture averages some 600 000 tonnes per annum, which is only 50 per cent capacity utilization of the cement industries. Due to the inefficient operation of the dust filters of the cement plants about 30 000 are released from the chimney as lime dust. This dust falls freely on the farms in the vicinity of plants, providing free soil conditioning lime to the local farmers.

Future prospects for lime consumption in Zimbabwe

The consumption of burnt limestone in the construction industry alone is expected to rise at some 10 per cent per annum. This will raise the local demand to some 9000 tonnes per annum by the year 2000.

The anticipated new construction and expansion of the existing lime consumers (sugar, metallurgy, glass, leather, agriculture and pulp and paper industries) are expected to consume a total of some 30 000 tonnes of lime per annum by the year 2000. This will represent an increased consumption of some 200 per cent within the next 15 years. This increase does not include demand for export. Therefore, during the period 1991 through to the year 2000, there will be an expansion of most of the lime consuming industries. By then, local consumption of lime will reach an estimated total of 40 000 tonnes per year. At the present rate of lime production, the existing installations will make available an estimated 18 000 tonnes per annum. Consequently, there will be a deficit of some 22 000 tonnes per annum and this quantity will have to be imported from abroad.

The planned expansion and rehabilitation of existing lime production plants by the year 2000 will allow a total production of some 50 000 tonnes per annum. Such a production will be adequate to meet the entire local market and leave an excess of some 10 000 tonnes per annum for the export market. Even if the local consumption of lime increases eightfold by the year 2000, full production, using already installed lime plants, will be adequate to demand. However, full production will require not only rehabilitation but also some modification to most of the existing installations. Of particular importance is the conversion from use of firewood to coal as fuel and overall energy conservation techniques.

Shaft kilns will remain the most appropriate during the period in question and there will be no need to convert to rotary or other types of kiln, even if it means sacrificing improved efficiency.

Some of the lime plants will need expansion to no more than 50 tonnes per day. By maintaining a small-scale level of production, it will be possible to have lime plants almost all over the country. This means that there is a greater chance that the industry will be able to fulfil the needs of nearby populations.

The shaft kilns and their small size will facilitate education of workers, maximize the use of skills found locally and avoid transportation problems in raw material supply as well as product distribution.

The total cost of rehabilitation and maintenance is estimated at Z$50 million at 1991 prices.

14. Basaltic rocks for cement and other binders

Ngo Van Minh*

In Vietnam, the demand for cement and other binders is high. There are four large-scale cement factories (of rotary kiln production technology) and 48 small-scale cement plants (of vertical kiln production technology), but the yearly average quantity of produced cement is only 50kg per person. This amount is low in comparison with other countries in the Asia-Pacific area such as in the Philippines (150kg per person) and in South

---

*Institute for Building Materials, Hanoi, Vietnam.

Korea (1000kg per person). In order to help meet the demand for cement and binders, the Government of Vietnam intends to develop various binders of low grade (such as lime-pozzolanic, lime-slag binders) and blended cements (such as pozzolanic cement, slag cement). The main advantages of these binders is simple production technology, locally available raw resources, widely used applications and low cost.

Basalt is one of the most popular pozzolanic materials of large reserve in Vietnam. Basaltic rocks are used as an admixture for cement and binders of low grade for producing mortar to bind bricks, tiles and low-cost elements.

Reactivity of basaltic rocks

Basaltic rocks (basalts) are situated particularly in longitudinal areas, from the 17th parallel to the south part of Vietnam. In the north, they occur at Phu qui, Vinh linh and Linh xuan. The basaltic rocks have different chemical properties (see Table 1).

The study of Hon nghe (Phu qui -Nghe tinh) basalt, which is one of the typical basaltic rocks, shows that this type of basalt has good pozzolanic reactivity. The lime absorption of friable basalt increases with increased burning temperature to 800°C. Its lime absorption has a maximum value at a burning temperature of below 800°C. The other dense metamorphic basalt has the highest degree of lime absorption after burning at a temperature of 500°C. At temperatures higher than 500°C, its lime absorption reduces. Pumice basalt has the same properties (see Table 2).

Basalt for low-grade binders

Due to good pozzolanic reactivity after appropriate thermal processing, the basalts can be used as admixture for low-grade binders and cement. The basalts of fine grinding (passing sieve of 4900 holes/cm^2) are mixed with lime with or without gypsum to create a binder with standard hardening times; and the strength of their mortar after 90 days is still gradually increasing. The products made of this binder, therefore, have increasing strength over time. Lime without basalt has very low strength (see Table 3).

As indicated in Table 3, the compressive strength of lime-basalt-gypsum binder is highest in comparison to the ones without gypsum.

Basalt as admixture for cement

Due to high pozzolanic reactivity, metamorphic basalts are good hydraulic admixtures for Portland cement. The experimental results show that the Bim son cement ground with 10, 15, 20 and 30 per cent Hon nghe basalt processed by burning at appropriate temperature does not change

Table 1. Chemical properties of basalts

No.	Type of basalt	SiO_2 (%)	Al_2O_3 (%)	Fe_2O_3 (%)	FeO (%)	TiO_2 (%)	CaO (%)	MgO (%)
1	Dense basalt	45.16	15.21	10.37	–	3.84	8.45	6.67
2	Metamorphic dense basalt	44.52	15.80	15.25	0.35	4.46	6.58	3.77
3	Pumice-basalt	44.35	15.57	10.08	2.2	3.81	8.48	6.47
4	Metamorphic pumice-basalt	42.90	16.46	13.28	–	3.10	6.56	3.77
5	Friable basalt	27.28	26.36	22.50	–	2.36	1.21	1.41

Table 2. Lime absorption of Hon nghe basalt and Son tay pozzolana

No.	Type of basalt	Lime absorption at temperature (mgCaO/g) (°C)				
		110	500	600	700	800
1	Metamorphic dense basalt	65.28	101.30	95.27	92.59	90.69
2	Metamorphic pumice-basalt	36.48	64.69	55.39	54.68	45.33
3	Friable basalt	50.66	56.08	61.80	71.10	81.28
4	Son tay pozzolana	90.04	96.43	106.32	153.60	181.40

Table 3. Physico-mechanical properties of lime-basalt binders

No.	Mix proportion (%)	Fineness (%)*	Density (g/l)	Hardening time Initial (min)	Hardening time End (h)	Water/cement ratio** (%)	Volume stability	Comp. strength of mortar (kg/cm²)*** 14 days	20 days	60 days	90 days
1	76 basalt + 20 lime + 4 gyp.	15.0	715	125	4.35	38.75	good	173	219	233	239
2	66 basalt + 30 lime + 4 gyp.	14.2	690	110	4.25	39.25	good	146	147	189	285
3	56 basalt + 40 lime + 4 gyp.	14.0	670	100	3.30	42.25	good	217	220	264	291
4	70 basalt + 30 lime	6.4	635	100	7.15	48.25	good	72	93	111	120

* Fineness: passing sieve of 4900 holes/cm².
** Water/cement ratio for mortar.
*** Compressive strength of mortar made of binder/sand ratio of 1:3.

Table 4. Physico-mechanical properties of Bim son cement ground with basalt

No.	Basalt proportion (%)	Fineness (%)	Volume stability	Water demand for cement paste (%)	Hardening time Initial (min)	Hardening time End (h)	Water/cement ratio for mortar	Workability (mm)	Comp. strength (kg/cm²) 3 days	7 days	28 days	60 days
X-00	0	12.4	good	23.00	165	3.30	0.46	117	111	169	320	337
X-10	10	12.6	good	24.25	145	3.30	0.46	118	116	175	319	389
X-15	15	13.6	good	25.00	175	3.55	0.47	117	110	194	335	371
X-20	20	13.6	good	25.50	150	3.45	0.47	117	104	161	301	345
X-30	30	13.2	good	26.75	125	3.50	0.47	117	103	152	280	311

Table 5. Compressive strength of blended cement mortar at age of 28 days (Ha tien Portland cement + Muarua basalt)

Specimen No.	Mixed proportion (%)	Compressive strength after 28 days (kg/cm^2)
XH-00	100% Portland cement (without basalt)	376
XH-10	90% Portland cement + 10% basalt	415
XH-15	85% Portland cement + 15% basalt	400
XH-20	80% Portland cement + 20% basalt	385
XH-30	70% Portland cement + 30% basalt	370

remarkably in its properties. The blended cement has a slightly higher water demand and a slightly longer initial hardening time than that of cement without admixture (see Table 4).

The results from Table 4 show that the blended cement with 10 to 15 per cent Hon nghe basalt has the equivalent or higher compressive strength at 28 days in comparison with Portland cement without admixture. The same results are achieved using blended cement made of Ha tien Portland cement and Muarua basalt originating from another location (see Table 5). This proves that all the tested metamorphic basaltic rock can be used as a hydraulic admixture for cement.

Conclusion

Metamorphic basaltic rocks of dense, pumice and friable state after processing at appropriate burning temperature have good pozzolanic reactivity. The appropriate burning temperature is 500°C for dense and pumice-basaltic rocks and 700–800°C for friable basalts.

Due to pozzolanic reactivity, all the tested types of metamorphic basaltic rocks can be ground and mixed with lime to become binders of low grade; and the lime-basalt binders can increase their strength remarkably, if they have 4 per cent gypsum added. The blended cement made of Portland cement and 10–15 per cent basalt maintains the standard properties in comparison with Portland cement without admixture. For instance, the compressive strength of the blended cement is equivalent to or higher than that of Portland cement without admixture. This can increase the output of cement plants by more than 10–15 per cent.

The results of this study indicate great potential for helping to solve the shortage of cement in Vietnam, by exploiting and using the large basaltic rock resource.

15. Chenkumbi Hills lime project

JOHN SPIROPOULOS*

Over the past ten years in Malawi, there have been up to 40 small producers of lime all employing traditional technology and operating largely on an informal basis. Most producers operate seasonally, producing only one or two batches per year between March and December.

Besides being seasonal, employment is casual and insecure. Typically,

---

*Formerly with ITDG, now Co-operative Planning and Education, Johannesburg.

wages are well below the government minimum and workers are paid only when owners' cashflows permit. There are approximately 900 people working in the lime industry during a typical year, of whom around 600 are working in the Chenkumbi Hills. It is estimated that between MK7000 and 10 000 (£1500–£2000) per year is paid out in wages in the area.

The main demand for lime in Malawi has been from the sugar industry. It consumes around 3200 tonnes of imported higher quality lime totalling about half the national consumption. The other important lime consumer is the construction industry.

The traditional technology is wasteful of fuelwood resources and the quality of lime produced is inadequate for the main consuming industry.

The primary aim of the Malawi government was to develop a local supply of high-grade lime to substitute for the imported material and save on foreign exchange. It was also concerned about the high fuel consumption of the traditional technology and the impact this has on the indigenous forest reserves.

Intermediate Technology (ITDG) was asked by the Ministry of Forestry and Natural Resources of Malawi to introduce improved lime production technology to the local small producers to upgrade the quality of lime produced and to cut fuel consumption.

Initial investigations showed clearly that poor kiln design was the main obstacle. The limestone available is a particularly difficult rock type. Although it is chemically of a high quality, it is a coarse crystalline marble which is difficult to calcine because of its friability on firing.

The objective of ITDG was to introduce and develop a technology which would raise the quality of lime produced and also reduce fuel consumption. This was to be done in a way which would increase or maintain the employment level, maximize income distribution, maximize sustainability through the use of local resources and skills, and stabilize the industry.

The strategic approach adopted by ITDG was to start with the traditional technology and through a process of identifying areas of improvement and, by progressively testing and modifying the technology, incrementally improve the quality of lime obtained and the fuel efficiency.

In practice, the project concluded at an early stage that the prevailing traditional technology could not be modified to make any significant improvement. A natural draught continuous operation vertical shaft kiln was constructed, tested, modified and tested again. This also failed to achieve the targets. Finally, a forced draught vertical shaft kiln, also continuously operated, was constructed and experimented with which did bring both the quality and fuel efficiency to a desirable level.

Raw materials

The limestone resources of the Chenkumbi Hills area are the largest single known deposit in Malawi. The measured reserves of the main limestone

Table 1.

Typical chemical analysis of Chenkumbi limestone	%
Calcium oxide (CaO)	52.06
Magnesium oxide (MgO)	2.23
Inerts and traces	2.69
Loss on ignition	43.02
Total	100.00

group stand at 3.7 million tonnes which is more than adequate for the foreseeable future.

At the start, fuelwood reserves were to be made available for lime production but the rapid influx of Mozambican refugees into the area resulted in a change of policy by the government. It was advised that the lime producers plant their own plantations. Alternatively they would have to use the commercial charcoal from Vipya plantation or the coal from the Northern region.

Traditional production

A traditional lime enterprise consists of a single owner performing the management function, a small core of permanent employees who do the more skilled and supervisory tasks and casual labour used wherever possible and paid on a piece-rate basis.

A typical kiln batch requires about 75 tonnes of limestone and uses locally available indigenous hardwood. The kiln design does not allow for the use of fast growing and burning plantation fuelwoods. Between 50 and 55 tonnes of wood is used per kiln batch to produce about 50 tonnes (2000 × 25kg sacks) of lime. The production cycle averages 60 days per batch.

The kiln is a rectangular box kiln constructed of limestone boulders cemented with lime-mud mortar. It has two firing openings at the base on the two short sides each leading into two trenches running along the length of the kiln.

The kiln is charged with alternate layers of fuel and limestone, ignited, stoked and then sealed and left to burn out and cool before discharging begins.

The quicklime is slaked manually. The hydrated lime is then sieved through a hand-punched sheet metal sieve to remove the coarse unburnt limestone. The product is milled in custom-operated modified maize mills and a small mineral processing hammermill. The bagging operation is performed manually either into three-ply paper sacks or into secondhand fertilizer or relief food sacks.

The quality of the lime produced using the traditional technology ranges between 32 and 45 per cent available lime. The thermal efficiency of the kiln is just below 15 per cent, i.e. only 15 per cent of the fuel consumed is used to calcine the limestone.

The problems faced by small producers are directly related to the technology used and the method of operation associated with it. While they have insufficient working capital and a generally low level of business skills, it became clear that the development and introduction of a suitable technology was the immediate priority. Without this the local industry would not be able to supply the market need, thus becoming marginalized.

The kiln technology was therefore prioritized for development, followed by the hydration, milling and sizing plant.

Natural draught vertical shaft kiln

The first, experimental kiln constructed at the Chenkumbi Hills was a simple natural draught kiln. The kiln shaft was 4.2 metres high and had an internal diameter of 1100mm.

Mild steel bar rings were to be built into the masonry to reinforce the kiln and prevent cracking. Unfortunately, due to bad supervision and work they were omitted. Consequently, with repeated firings, the kiln started to crack.

After the first firing trial, the kiln was modified by extending its length to 6 metres and gradually narrowing the internal diameter of the shaft from 1100mm at one metre from the top to 800mm diameter at the top.

Quarried limestone was dressed in the quarry area to kiln feed size (75–125mm) and charged directly into the top of the kiln via a timber charging ramp.

The first firing trials in the simple vertical shaft kiln were with fast-growing plantation fuelwood to test its usefulness as an alternative to the hardwood used traditionally.

The firing temperature reached in the hottest zone was 950–1050°C. The firing zone in the kiln was between 4 and 4.5m out of a total kiln height of 6m. This was due to decrepitation of the rock on firing. The result was channelling of air, poor heat distribution, poor fuel use and insufficient cooling.

The quicklime was discharged and slaked manually during the day shift. The slightly moist and hot slaked lime was left to dry out and cool for 24 hours. It was then sieved, milled and bagged in the traditional manner.

Approximately 9 tonnes of limestone were quarried to produce 7.2 tonnes of kiln feed. The charging rate was 300kg limestone and 120kg fuelwood per hour. The discharge rate of quicklime was about 220kg per hour. 150–155 of the 660 litres was used to produce 5 tonnes, i.e. 200×25kg sacks of lime hydrate per day.

The quality of lime produced ranged between 42 and 48 per cent available lime out of a possible 69 per cent. The product contained more fine material than that produced with traditional technology.

The fuel used was the fast growing plantation eucalyptus. The thermal efficiency of the kiln was calculated to be 30 per cent. This could be improved to 36 per cent if the fuelwood were air-dried before use.

The natural draught kiln with improvements failed to interest the lime producers. An informal survey of opinion indicated that, although producers were impressed with the ability of the new kiln to use plantation fuelwood, their dominating ambition was to produce for the high quality market. Clearly, the new kiln could not do this.

Also, the new kiln constituted a considerable investment and a higher risk. None of the producers had direct experience in operating vertical shaft kilns. They were therefore uncertain whether they could achieve the improvements obtained during the trials by ITDG.

To the extent that no interest to take up the new kiln emerged, the technology and the strategy adopted can be considered a failure. The experience, however, provided essential information on the behaviour of the rock on firing and gave clear directions for the next steps in research and development.

Experimental forced draught kiln, Chenkumbi

In this system, air is forced into the kiln under pressure through a tuyere at the base of the shaft. It was felt that this type of system would provide the necessary measure of control over the firing conditions, i.e. length of the firing zone, temperature and heat distribution. In addition, it is relatively inexpensive and robust compared to induced draught systems which draw air through the kiln from the top.

The first objective of these trials was to show that the kiln could operate efficiently using charcoal and coal as fuels. The second was to prove that the technology could produce a suitable quality lime for the sugar industry.

The limestone feed size was between 75 and 125mm, and the fuel used in the trials was charcoal with a 85 per cent fixed carbon content and a net calorific value of 30MJ/kg.

The kiln used for the trials was the natural draught kiln described previously, modified to incorporate the forced air system.

A centrifugal fan blower with a rated capacity of 1000m^3/hr against 20 inches (500mm) water gauge provided the air supply and was driven by a small diesel engine. The blower was over-designed.

Firing trials

The variables that determine kiln operating efficiency are as follows:

- kiln design including shaft height, diameter, and height to diameter ratio
- materials use in kiln construction, especially the insulation

- limestone and fuel type and quality, and kiln feed size and size distribution
- limestone to fuel ratio
- air flow through the kiln
- frequency of discharge.

The operating conditions that were varied to change the performance of the kiln were as follows:

1. Frequency of discharge was used to vary the position of the firing zone and also the retention time of the limestone in the kiln. The same amount of lime was discharged from each discharge door at the set discharge rate. Each discharge was between 220 and 240kg. The charging rate was determined by the rate of discharging. The kiln was fed when a space developed after a discharge.
2. The air flow determines the rate of combustion of the fuel and, therefore, the length of the firing zone and the temperature. Air flow was controlled by means of a damper at the fan inlet and a variable bleed at the discharge side of the fan. It was measured indirectly by recording the pressure drop across an orifice plate in the ducting connecting the fan to the kiln. A simple manometer fixed to a board consisting of a plastic tube with coloured water in it and a school ruler were used to measure the pressure difference in millimetres. A calibration curve of pressure drop (mm) against air flow was prepared and used.
3. The stone to fuel ratio was varied to change the temperature in order to achieve optimum fuel efficiency and calcination.

A sample of burnt lime which had passed through the firing zone at stable conditions was taken, slaked and analysed on site.

Results

Six trials were carried out over a period of 300 hours. Table 3 (page 147) summarizes the conditions and results for each trial. Trial 6b was conducted with a 50:50 wood:charcoal mix as fuel. All other trials were fuelled with charcoal only.

The firing zone was located at the centre of the shaft and the temperatures were set at 1000–1050°C.

Table 3 shows a gradual improvement through the trials, the most noticeable improvement being due to reduction of the discharge rate to 105 minutes. This corresponds to a charging rate of 240kg/hr, which is somewhat higher than expected (150–200kg/hr). The discrepancy is probably due to decrepitation of the rock on firing resulting in compaction.

The optimal discharge rate for the given fixed variables is somewhere between one hour 30 minutes and one hour 45 minutes.

The amount of excess air being put through the kiln mass was much higher than expected, although the measured air flows should be treated with some caution given the difficulties in achieving accuracy. The excess air was measured to be around 100 per cent in the best trials. The main reason for this was air channelling its way up the sides of the shaft, noticeable especially during the last trial with wood where temperatures at the centre of the top of the kiln were much higher than close to the walls. Smoke (unburnt volatiles) was seen to come from the centre.

In the final analysis, high excess air rates are not important provided the efficiencies are high.

A discharge every one hour 45 minutes produces approximately 140kg output (CaO) per hour. This is 3360kg in 24 hours. The amount of limestone charged into the kiln was about 5.76 tonnes and the amount of waste screened after slaking was about 500kg per day.

The CaO product was, therefore, in the region of 3000kg per day. Approximately 1.4 tonnes of water was used per day and 795kg of this combined with the CaO to produce the lime hydrate. The remainder evaporated in slaking. Approximately 150×25kg bags of product were produced per day.

The decrepitation is inherent to the rock type and is discussed elsewhere. Its effect is the production of two products. The outer surface of the rock is thoroughly calcined, breaks off, and forms a high quality product (fines). The inner core of the rock is only partially calcined or remains uncalcined. This is the second, lower quality, product (see Table 3).

Using a manual method of slaking described previously, a conservative estimate suggests that around 30 per cent, and possibly more, of the lime produced would meet the high quality requirements of the sugar industry set at a minimum of 60 per cent available lime (see Table 3).

The stone:fuel ratio was increased from 5.9 to 7.1. The amount of fuel (charcoal) used was 34kg per hour, which is equivalent to 816kg per day. The thermal efficiency was calculated to be 36 per cent.

A heat balance calculation indicated that less than 10 per cent of heat losses were in the exhaust gases and discharged lime. The exhaust gas heat losses are usually an important indicator of efficiency. During the trials gas temperatures remained remarkably low at around 60°C. There is, however, a discrepancy of between 40 and 45 per cent. This constitutes heat losses through the kiln walls which are higher than usual probably due to heat loss through the cracks that appeared in the kiln and through the inspection holes.

The experience with the experimental forced draught kiln at Chenkumbi Hills was a very positive one. The lime producers showed an immediate interest in it. One of the producers promptly built his own vertical shaft structure.

We showed that suitable quality lime for the sugar industry could be produced by this kiln – based on the sugar corporation's own analyses. In addition, we were now in a position to make modifications to the kiln to improve its efficiency and its durability.

Balaka forced air kiln

The next stage of the research and development effort was to develop a low cost hydration, milling and classification plant. We approached a local USAID-funded enterprise development organization for financial support on this. They recognized the potential and agreed to assist.

A consortium of local business people was brought together to participate to taking over the project once the R&D phase was completed. A joint venture was entered into with USAID providing financial and management input and ITDG providing the necessary technology development skills.

The slightly modified forced draught kiln was constructed in Balaka. Limestone was quarried from a site in Chenkumbi Hills by means of drilling, blasting and then manual breaking and dressing to kiln feed size. The kiln feed was then taken to the Balaka site by hired trucks.

The kiln was constructed with the same internal dimensions as the experimental kiln at Chenkumbi, i.e. an effective shaft height of 6 metres and 1.1 metre internal diameter tapering one metre from the top to 800mm diameter.

The kiln was lined with locally produced refractory bricks. These were custom-made for the first time in Malawi specifically for the emerging lime industry. The refractories were wedge-shaped fired clay bricks, in two sizes, to facilitate tapering of the shaft at the top.

Outside the refractory lining, there was a 230mm layer of insulation bricks, once again made specifically for the purpose, and outside of that was a 120mm skin of ordinary clay stock brick. The brickwork was encased in a 6mm mild steel casing with a 2.3m diameter.

The kiln is free standing. A steel platform with wooden decking was built on the top of the kiln. The kiln charge was hoisted by means of buckets drawn up by windlass fixed to a davit arm. The davit arm can swivel from over a charging hopper at the centre of the kiln.

The temperature in the kiln was set at between 1000 and 1100°C and checked at the commissioning stage by means of thermocouples. Later, once the operation stabilized, it was possible for the kiln operators to judge the temperature fairly accurately by eye.

Initially there was some difficulty in balancing air flow with retention time and discharge rate. While this kiln was built with the same dimensions as the kiln at Chenkumbi Hills, it did not perform in the same way. But, within four or five days from start-up, it was possible to establish visually that the product from the kiln was being well calcined. There was no sign of charcoal in the discharge and a relatively low quantity of core (uncalcined limestone).

The kiln fan was installed with a horizontal bottom discharge mounted on a concrete plinth above ground level and driven by a 5.5kW motor. Air was fed into the kiln by the fan through a 150mm pipe to a series of cast-iron tuyeres forming a manifold in the middle of the kiln.

The fan was calibrated using a pitot static tube to measure flow rates, and a manometer to measure pressure rise. The fan was set to give 650m^3/hr, to match the optimal flow rate observed during the Chenkumbi trials. In Balaka, at 650m^3/hr, the pressure drop was 20–25mm water gauge. This seems very low, but it correlates well with figures given by Wingate.[5] Although there is no comparative figure for the Chenkumbi Hills kiln, the indications from the electrical current drawn are that the pressure drop there was higher.

The kiln discharge rate was set at every 90 minutes, compared with 105 minutes at Chenkumbi. This increased the output of the kiln and therefore also the amount charged into it.

The amount charged into the kiln was 5720kg limestone. There was a loss on ignition of 36–37 per cent, i.e. about 86 per cent conversion of the $CaCO_3$, resulting in a CaO output of around 3400kg per 24 hour day.

Approximately 970kg water is added to this in 24 hours to produce 4370kg lime hydrate including waste. From this about 11 per cent is sieved off as plus 5mm waste which leaves material for milling, classification and bagging of 3890kg.

The lime produced averaged at around 60 per cent available lime. The particle size distribution, after slaking and sieving, was shown to be 70 per cent less than 150 microns. This quality is within the specification of the sugar corporation. The bulk of the imported material is of a quality well below this. The selling price, ex-works, for this higher quality lime is nearly double the price to the lower quality building market.

The fuel that can be used by this kiln is charcoal (hardwood or softwood), coal or fuelwood. Fuelwood was not tried, however, because its use is not considered a long-term prospect for lime producers.

The fuel consumption was found to be 792kg charcoal or coal per day. The kiln thermal efficiency for the Balaka plant was 42 per cent compared to 36 per cent for the experimental kiln. The difference of around 6 per cent is due primarily to the insulation built into the walls of the Balaka kiln.

Conclusions

Kiln efficiency and quality

The change from batch to continuous operation, i.e. from the traditional kiln to the natural draught vertical shaft kiln, resulted in a doubling of kiln efficiency and a saving in fuel cost of 42 per cent. It also resulted in almost 9 per cent increase in the fractional conversion of the limestone, that is a 9 per cent improvement in quality.

The Chenkumbi Hills natural draught kiln was then modified to a forced draught system which resulted in a 7 per cent increase in kiln efficiency only. The increase is predictably less dramatic.

However, the forced draught kiln at Balaka, which has the same dimensions as the Chenkumbi experimental kiln but includes an insulation in the kiln wall, is an additional 6 per cent more fuel efficient.

In addition to the improvements in kiln efficiency, the forced air system raises the limestone conversion from 63 to 86 per cent. This is a substantial improvement in quality leading to the substitution of imported lime and a foreign exchange saving of up to 1.26 million Malawi Kwacha (approx. £250 000) per annum.

Another clear advantage of the vertical shaft technology is that it can use fuels other than indigenous hardwood. This alleviates the pressure on an already overstressed natural environment.

Finally, the kiln efficiency obtained by the Balaka forced draught kiln, which could probably be improved further by even better insulation, is compared below with efficiencies of other modern kiln designs (Table 2, figures obtained from Boynton[1]).

Table 2.

Balaka forced draught kiln	42%
(100% $CaCO_3$ conversion – see Table 4)	
Conventional rotary kiln without heat recuperation equipment	35%
Conventional rotary kiln with heat recuperation equipment	50%
Kiln with highest kiln efficiency on record	85%

Investment, income and employment

There is a large increase in the estimated investment cost from MK15 000 for the traditional operation to MK428 050 for the forced draught operation with hydrator, and classification and milling plant. The increase in annual returns is also considerable. The Return on Investment, however, is not substantially increased at all (see Table 6).

A potential investor would have to consider the comparatively small margin in the Return on Investment and the added management effort required to run the forced air operation against the increased annual returns. The risk is higher for the forced air operation but the increased returns may justify investment.

At the macro level, government would have to recognize the advantages that adoption of the improved technology represents in import substitution and foreign exchange savings, and provide the necessary incentives. These might be indirect through, for example, provision of adequate infrastructure or more directly through tax holidays, low interest finance, tariffs on imported lime, and so on.

It should be noted that the capital cost of the traditional operation, shown in Table 6, does not include the milling component of the process. It has been assumed that this is a separate enterprise, as is the case in the Chenkumbi Hills area where there are four custom millers processing lime for nearly 20 producers. It has also been assumed that none of the existing producers will be able to secure any loan finance. The income statement, therefore, does not include depreciation or loan repayment figures.

It should be noted in the comparative income statements (Table 6) for the three levels of technology that the wage bill increases with increased sophistication while the fuel cost diminishes.

On the other hand, the income statement for the forced air system shows a comparatively high loan repayment component. This points to higher risk and with it the need for more efficient management practices.

The first benefit of the vertical shaft technology is greater productivity. Secondly, the shift towards greater sophistication leads to a higher wage bill. In relation to this it should also be noted that the vast majority of workplaces in traditional production enterprises are casual, insecure and erratic. The traditional enterprises distribute income more widely while the vertical shaft options concentrate labour and income distribution. From one perspective this might be viewed negatively, but from another, in a community of around 2500 people, distribution to 400 people might be adequate. Adoption of the more advanced technology will foster local enterprise development and wealth accumulation. Theoretically this may lead to greater investment in additional productive enterprises.

Technical details

The kiln construction and the materials used varied from the rudimentary box-type structure made of limestone boulders and mud mortar to the relatively sophisticated shaft kiln with steel casing. While the rough box kiln is not suitable as a design, for the reasons expressed later, use of limestone boulders as a construction material is quite acceptable in principle. It is locally available, requires no fuel for processing and it is cheap.

The experience of this project shows that if local materials, e.g. limestone boulders, are used, the investor must make sure the masonry skills are of sufficiently high quality to obtain a well-built kiln casing. The kiln should have steel straps at 400mm centres and 150mm from the top of the shaft.

Other casing options are reinforced concrete or steel. It should be noted that steel casing is only preferred for its ease of erection, otherwise reinforced concrete or brickwork and stone masonry with steel strapping are quite strong enough.

The dimensions of the kiln could be modified. In principle, the heat distribution in the kiln and therefore also the quality will improve if the diameter is larger. The height of the kiln could also be increased to allow for a longer firing zone and a more gradual temperature profile in the kiln.

The kiln dimensions recommended for a new custom plant planned for implementation at Chenkumbi are 1.2m diameter and 7.2m height. The kiln construction will not have a steel casing as at Balaka but will be a fired clay brick structure with steel strapping.

The fan used at the Chenkumbi Hills experimental kiln, which was later transferred to Balaka, was found to be considerably overspecified at 1000m^3/hr through a 5000Pa pressure rise. A 800m^3/hr through 500Pa pressure rise, overhung, centrifugal fan, V-belt driven by a 4-pole 1.5kW electric motor, has been specified for the proposed new plant at Chenkumbi Hills.

Choice of technology – the issues and criteria

The following are the types of conditions and factors which weigh in favour of small-scale kilns.

The first situation favouring small-scale plant is widely dispersed, small pockets of demand for lime where the distances between production centres and locations of demand are high.

Secondly, a single factory such as a sugar refinery or a metallurgical plant might wish to secure a consistent supply of lime. In this instance, it could set up its own small-scale plant.

Another condition in favour of small plant is where only small deposits are available which cannot justify a large-scale investment.

Finally, the environment might be such that wage rates are low, unemployment high making rural industries development a priority. In these circumstances, a small plant can, as has been shown by the experience in Malawi, maximize distribution of income, employment and wealth, and still produce a high quality product.

Table 3.

Trial number	Charge rate (kg/hr)	Discharge frequency (hrs)	Stone:fuel ratio	Fuel (kg/hr)	Air (mm)	Air flow (m³/hr)	Stoic air (m³/hr)	% excess air	Available Mixed	Lime Cores	Fines	$F_c{}^1$	Kiln efficiency[2] (%)
1	233	1.00	5.9	40	32	700	382	83	48			0.62	22
2	209	1.50	5.9	35	31	800	342	134		57	61	0.77	27
3	205	1.50	5.9	35	18	650	342	90	47			0.61	22
4	253	1.50	5.9	43	25	740	412	79	54			0.70	25
5a	326	1.50	7.1	36	32	810	440	84	45			0.58	25
5b	275	1.50	7.1	39	18	650	371	75		58	60	0.76	34
5c	275	1.50	7.1	39	18	650	371	75		57	59	0.75	33
6a	241	1.75	7.1	34	18	650	325	100		59	66	0.82	36
6b	245	1.75	5.3	47	18	650	322	102		47	66	0.75	33[3]

Notes:
1 F_c is the fraction of calcium carbonate ($CaCO_3$) that has been calcined in the kiln. Its calculation is based on 10 per cent unburnt cores and 40:60 core:fines ratio.
2 Kiln efficiency is the proportion of heat (calorific value) that has been used for calcination. The calculation assumes a heat of calcination is 2740KJ/kg. The figures show the fractional efficiency, i.e. they are corrected for the fraction of the calcium carbonate converted to lime.
3 Efficiency of trial 6b based on wood charcoal mix.

Table 4. Fuel consumption and efficiency

	Traditional kiln	Natural draught kiln	Forced draught kiln – Chenkumbi	Forced draught kiln – Balaka
Limestone charge per day	75 tonnes/ batch	7200kg	5500kg	5500kg
Fuel charge per day	55 tonnes/ batch	2880kg	816kg	792kg
CaO produced	42 tonnes/ batch	4280kg	3000kg	3400kg
Fraction of $CaCO_3$ converted to CaO	0.58	0.68	0.82	0.86
Fuel used per tonne CaO produced (MJ)	19640	10090	8160	6928
Calorific value of fuel (MJ/kg)	15	15	30	30
Fuel cost per tonne lime produced (MK)	94,00	54,00	40,80	34,95
Kiln efficiency (%) – assuming 100% conversion	15	29	36	42
– assuming fractional conversion	9	18	29	36

* CaO content of limestone 91.4%.

Kiln or thermal efficiency is the ratio of heat used in calcination to total heat consumed. It ranges from 15 per cent for the simplest wood-fired batch kiln to 85 per cent for the most efficient modern kilns. Efficiency is calculated using the following formula:

$$\% \text{ kiln efficiency} = \frac{\Delta H_c \times L_s}{C_f \times M_f/M_p}$$

where:

ΔH_c is the heat of calcination per tonne CaO produced (theoretical heat required). This is 3200 MJ/tonne CaO.
L_s is the available lime content of limestone.
C_f is the calorific value of the fuel.
M_f/M_p is the amount of fuel per tonne CaO produced, in kg.

The formula assumes 100 per cent conversion of the limestone to lime. A more accurate reflection of efficiency would be fractional conversion, i.e. full conversion kiln efficiency multiplied by 'Fraction of $CaCO_3$ converted to CaO' factor (see above).

Table 5. Income statement

	Traditional production	Natural draught production[8]	Forced draught production[9]
Sales (2000 × 25kg bags lime)	10 000	10 000	17 000[6]
Production costs			
Labour cost[2]	930	1 005	1 640
Blasting cost	–	475	470
Fuel cost	3 960	2 325	1 535
Bags (2000 bags @ 50t)	1 000	1 000	1 000
Maintenance costs	3 000[4]	1 180	1 880
Production supervision	180	300	500
Total production costs per 2000 bags	9 070	6 285	7 025
Overhead costs	455[1]	940[3]	1 055[3]
Loan repayment[5]	–	1 440	5 375
Depreciation costs	–	270	1 930
Total costs per 2000 bags lime	9 525	8 935	15 385
Gross profit per 2000 bags lime	475	1 065	1 615
Production per annum (tonnes $Ca(OH)_2$)	150	1 050	1 065

Notes:
1 Overheads calculated on basis of production of three batches per year (5% of production costs).
2 Daily government minimum wage rate used in calculation.
3 Overhead set at 15% because of continuous production.
4 Milling cost charged at MK1.50 per bag.
5 Repaid over 5 years in equal instalments. The interest rate is 20% (estimate). Calculated using @PMT function on Lotus 123 spreadsheet program.
6 660 bags at MK5.00 and 1400 bags at MK10.00.
7 Charcoal cost per tonne is MK150.00 delivered to site.
8 210 working days (due to additional repair and maintenance).
9 275 working days.

Table 6. Capital costs

(All figures in Malawi kwacha)	Traditional box kiln	Natural draught vertical shaft kiln	Forced draught[3] vertical shaft kiln
Kiln	500	10000[1]	59000[4]
Hydration, classification and milling	–	70000	229500
Buildings	4500	6500	81000
Sub-total	5000	86000	379000
Working Capital	10000	26620[2]	48150
TOTAL INVESTMENT	15000	113120	428150
ANNUAL RETURN	1425	21420	58820
ANNUAL RETURN ON INVESTMENT	9.5%	18.9%	13.7%
Depreciation cost (per annum)[5]		5600	23170
Loan finance requirement (assumed)		90500	342500

Notes:
Five Malawi Kwacha is equivalent to one Pound sterling (approximately).
1 Cost of kiln includes all materials – steel straps, chimney hood, ramp and kiln. Kiln lining is traditionally of fired brick which would be totally replaced every six weeks (one week to replace). Cost also includes labour and supervision. No amount included for commissioning or training.
2 Two months production cost.
3 Based on revised plant cost estimates.
4 Kiln including installation and commissioning.
5 Plant and equipment depreciated on a straight line basis with no salvage value over 15 years, and building in the same way over 25 years.

16. Refractory bricks for lime kilns: small-scale production using local raw materials

CHRISTOPHER STEVENS*

To date, Dedza Pottery has made the bricks for four vertical lime kilns in Malawi. The first was in 1988 when we were asked to make refractory bricks for a lime kiln in the north of Malawi. Knowing little about lime production or vertical shaft kilns at that time, we made the bricks to the special shape requested but using our normal refractory brick mixture and moulding methods. In fact this kiln, no. 1, proved to be reasonably sound and long lasting as, by chance, we had satisfied many of the requirements that we later found were necessary for good lime kiln refractories.

*Dedza Pottery, Malawi.

In 1989, we were asked to produce bricks for kiln no. 2. This was being designed with assistance from ITDG with the aim of eventually producing very high quality lime so that the large sugar industry in Malawi would be able to buy locally rather than import. A complete factory for burning and processing the lime was being built and the very best refractories were requested. We spent six months carrying out tests to make bricks appropriate to the special needs of lime kilns.

After trying many different raw materials, in dozens of different combinations and trying several different forming processes, we concluded that the mixtures and methods used to make the bricks for kiln no. 1 could only be economically upgraded by the additional procedure of allowing the mixture to mature slightly longer before moulding the bricks.

The testing had, however, given us many ideas for further research. Six months after commissioning kiln no. 2, there was a gross overfiring when the fuel was changed from charcoal to coal; some of the refractory bricks actually melted to glass. Before the kiln was repaired, I had the chance to inspect the damage caused by the overfiring and also the wearing of the bricks in all parts of the kiln. Much was learned from this inspection.

During 1990, having been advised that new kilns would be constructed in 1991, we continued our research and finally came up with a much improved brick which has been used to build kiln no. 3. A further improvement was then made during 1991 and these bricks are about to be built into kiln no. 4. At the time of writing, kiln no. 3 has not yet been fired.

The challenge

Based on experience to date, the ideal lime kiln brick would have to be:

- refractory enough to withstand day to-day working temperatures and some degree of overfiring. It should not spall or crack with sudden heat applications.
- resistant to lime and the fluxing action of hot quicklime.
- resistant to abrasion so the limestones do not wear the bricks out too fast.
- available to uniform standard sizes.
- economically priced and available not too far from the kiln site to avoid large transport costs.

Added to this, for small-scale production in non-industrialized countries where demand for such specialized bricks will be small and intermittent, they should be made by a process where any expensive capital equipment will have alternative uses and any specialized equipment specifically for making lime kiln bricks, e.g. moulds, will have a comparatively low capital cost.

The final type of brick that should be produced in any situation will be a compromise of the above, often conflicting, specifications. Below are some notes regarding each of the above specifications and how they can be achieved, and the compromises actually made in production.

Refractoriness

The working temperature at which a firebrick can be used depends to a very large extent on the raw materials it is made from. Many fireclays are usable 'as dug' with only slaking or grinding to break up lumps. Impurities that are in the form of sand, stones or roots, etc., can be removed by slaking the clay (mixing with an excess of water) and wet sieving. This is a process that is commonly used for clays for pottery but may prove too expensive for bricks. It also requires large tanks and drying areas. Otherwise, there is very little that can be done easily to improve the refractoriness of a raw material other than combine it with another raw material with a much higher refractoriness.

A fairly high firing plastic clay is essential for making refractories using less industrialized processes. For small-scale production, deposits that might not be considered adequate from a Geological Survey, and therefore not be recorded, might be suitable. Often white or grey clays are a good indication of refractoriness and can be located by noting where village houses have been 'whitewashed' with a white clay. If the clay is white or light pink when fired to 900°C, and is fragile at this temperature, this is also an indication that further testing should be done. When fired to its proposed working temperature, the clay should still be able to absorb over 15 per cent of its own weight when soaked in water.

Refractoriness is usually measured by 'pyrometric cone equivalent' or PCE. This is the number of the standard pyrometric cone that begins to melt at the same temperature as the material under test. Commercial firebricks will normally be rated by their PCE. PCEs have temperature equivalents though the temperature varies slightly according to kiln atmosphere and rate of firing. For lime kilns it is suggested that the PCE of the refractory bricks should be 270°C–300°C higher than the maximum expected working temperature. For example if the normal working temperature of a lime kiln is expected to be 1200°C, and 100°C extra is allowed for possible overfiring, then the maximum working temperature of 1300°C would require a brick that begins to melt at no less than 1570°C which is a PCE of Cone 26. In refractory terms this is considered a 'Low Duty' brick. Determination of such high PCE needs to be done in a specialized laboratory, though as only 500g of the material is needed for testing, it is perfectly possible to send the clay abroad for the test. 1991 prices were US$100–200 per test.

If a chemical analysis of a clay is available, then clays with over 30 per cent Al_2O_3 and less than 5 per cent in total of Fe_2O_3, CaO, MgO, Na_2O and K_2O would probably have a PCE of at least Cone 26.

The PCE increases in proportion to the amount of Al_2O_3; 40 per cent Al_2O_3 would probably give a PCE of Cone 33 or 34. There are a few raw materials that have a very high Al_2O_3 content and these can be added, suitably ground, to the clay mixture to greatly improve refractoriness. White or grey bauxite, sillimanite and kyanite are the ones most likely to be found. The cost of mining, transporting and grinding will have to be balanced against the improvement in the bricks.

Additions of quartz sand, which is often more easily found than the above raw materials, will also increase the refractoriness of a clay. However, due to the large expansion of free silica at about 600°C, it is best avoided as the bricks are more likely to crack on heating and cooling. There is also an additional problem with free silica as mentioned below.

Resistance to lime

Another pitfall for lime kiln refractory brick makers is the existence of 'eutectics'. This is the principle whereby two or more ingredients, which each individually have very high melting points, when combined in certain proportions will produce a melt at a much lower temperature. Of particular concern are the following:

Silica oxide has a melting point of 1620°C
Aluminium oxide has a melting point of 2050°C
Calcium oxide has a melting point of 2580°C

Yet these three oxides (all present in a working lime kiln) can melt at a temperature as low as 1170°C. The dangerous combination is a high silica and low alumina one, so this is an additional reason for increasing the alumina and not the silica content of the bricks.

Abrasion resistance

Clays become tough and abrasion-resistant when they begin to fuse as the temperature approaches their melting point. Thus the two requirements of refractoriness and abrasion resistance are conflicting.

It was noted from the inspection of kiln no. 2 that the greatest abrasion occurred in the top 30 per cent of the kiln where the fuel and lump limestone is loaded. I was advised that the maximum recorded temperature in this part of the kiln was 800°C. Thus for kiln no. 3, we have made the top bricks from a high firing brick clay, fired to 1100°C, at which temperature the bricks have become fairly tough; however, firing too near the maturing temperature of a clay has its own problems, as small variations in kiln temperature produce considerable differences in the sizes of the finished product. This resulted in a considerable number of discarded bricks and more size variation in the delivered bricks.

One of the many tests we conducted before making the bricks for kiln no. 2 was to try grinding the clay mixture finer. It appeared to have little effect until the size of the grind reached 250 microns, at which stage a remarkable improvement in abrasion resistance was recorded. However, to grind all the mixture to this size would be very expensive, so a composite brick was tried with only the wearing end of the brick made from the expensive finely ground mixture. This proved excellent under test conditions but losses of over 80 per cent in a trial production meant that it could not at that time be used for production. However, a few of the bricks that survived were included in kiln no. 2. The result was quite dramatic; when I inspected the kiln after six months of use, the composite bricks were protruding from 10mm to 15mm beyond the ordinary refractories, giving them an expected life of at least twice the ordinary refractories. This year, we hope we have overcome the problems in making the composite bricks and all the bricks we are completing for kiln no. 4 are of this type. They cost about 25 per cent more to make than ordinary refractories.

Standard sizes

Dedza Pottery now produces two sizes of tapered brick. Using different combinations of these two bricks, it is possible to build kilns with diameters anywhere between 1000mm and 1500mm (internal), including the taper at the top.

The size is: length 220mm, height 108mm, large face 100mm wide: small face 65mm wide (small brick) and 75mm wide (large brick).

Appropriate production techniques

Our experiments with dry pressing, as is used industrially, showed that with our raw materials a press capable of producing 20–50 tonnes would be necessary to produce a similar density of bricks as could be obtained by moulding by hand, using metal moulds and fairly stiff plastic clay. We found that the maximum shrinkage from mould size to fired brick should not exceed 7 per cent and 5 per cent was ideal. Higher shrinkages cause cracking problems both in drying and firing. Most firebricks are made using a 'grog' of the same clay pre-fired and ground up and added to the mixture. Originally we used a low-fired (900°C) grog which was easy to make and soft to grind. This solved the cracking problems during drying but the shrinkage when the brick was fired to its final temperature remained unchanged. This only became a problem when we tried to make the composite bricks. We finally found that, for these, the grog must be fired to the same temperature as the final brick; thus the firing is more expensive and the regrinding is considerably more difficult.

The amount of grog required can only be determined by testing; it is expensive so as little as possible is used to produce good bricks. A starting point might be 50 per cent by volume ground clay to 50 per cent grog. If the high alumina raw materials mentioned previously are available, these can be used instead of some or all of the grog.

It is possible to grind by hand but some mechanical assistance is usually more economic even where labour is comparatively cheap. We use hammers, roller grinder, hammermill and ball mill for the different parts of our composite bricks. All these machines are available from our other ceramic production. Suggested grog sizes are 2mm for main body and 150 microns for tip.

The most essential piece of equipment is a kiln for firing the bricks, up to 1300°C if possible. Fuel burning kilns are very inefficient in small sizes; our kiln that fires 1500 lime kiln bricks uses less than twice the fuel of one that fires only 300 bricks.

Economics

Before deciding on exactly the type of bricks to be manufactured, all the economics of lime production, downtime of the kiln while it is relined, etc., need to be considered. In most cases better quality and hence longer lasting bricks will eventually repay the initial extra outlay.

Concluding caution

Some of the advice given above, particularly on composite bricks, is untested and can only be confirmed after both kilns 3 and 4 have been fired for some time and then inspected.

I would always advocate a cautious approach in ceramics; we do not fully understand what happens in high-temperature firings and whether it is clay floor tiles, new glazes, teapot spouts or lime kiln bricks, there will usually be 50 disappointments for one success. This gives a particular excitement when something eventually seems to be coming out as planned. It does appear, however, that our lime kiln bricks are coming out even better than planned.

Future plans

We hope to develop a method of determining PCE at high temperatures by firing no higher than Cone 10. This will enable PCE testing to be done locally, cheaply and quickly.

17. Tanzania – small industries development of lime kilns

E G S IKOMBA*

Super Amboni Lime Products is a privately owned company which produces lime for the building industry and various industrial uses. The TSh20-million project has been financed through a loan from the National Bank of Commerce (NBC) and constructed under the technical guidance of the Small Industries Development Organization (SIDO), which is the main adviser on construction, commissioning and general troubleshooting.

Most of the fittings used in the kiln have been fabricated locally by SIDO's Common Facility Workshop in Tanga Region.

Project description

The project is located at Amboni Village which is 8km from Tanga Municipality and 357km from Dar es Salaam. It has all the basic services, such as tap water, electricity, and a tar road.

The project will have three lines of products: lime for building and industrial use, precipitated calcium carbonate and sand lime bricks (calcium silicate bricks).

At present it is dealing with the production of lime for various uses.

Raw materials

The project has 17 hectares of deep limestone deposit under a 99-year lease from the Ministry of Lands. Limestone samples from this quarry have been analysed with the following results:

SiO_2	0.44%
Fe_2O_3	0.08%
Al_2O_3	0.26%
CaO	55.04%
MgO	0.31%
LOI	43.41%
Total	99.54%

The quarrying of limestone is done by blasting and then casual labourers break it manually with stone hammers to the required size of 12cm before

---

*SIDO, Dar es Salaam.

the stones are fed into the kiln. Sometimes stones are broken (sized) using a nearby crusher belonging to another entrepreneur.

Kiln design

The project's kiln has a rectangular shape which was constructed using cement bricks having a ratio of 3:1 sand:cement. The lining of the kiln has been constructed with firebricks which were formerly used by Tanga Cement Factory.

This kiln, with internal dimensions of 5.3 × 4m, was constructed under the guidance of SIDO. The design provides a high level gallery giving access to firing chambers on two opposite sides of the square shaft. The height of the kiln is 15 metres and has the following features:

- For controlling temperature, a thermocouple read every four hours, in conjunction with batch release, has been fixed.
- At full capacity, a discharge approximately every four hours should produce 20 tonnes per batch.
- The kiln capacity is 40 tonnes per day.
- Approximately 10 per cent underburnt lime is fed back to the kiln for re-burning.

The fuel used for burning limestone is industrial diesel oil (IDO). The kiln uses 2000 litres per 24 hours operation. A litre of IDO costs TSh40 and it is readily available from Tanga Municipality.

Method of operation

The kiln is equipped with one service tank and one reserve tank. Oil is pumped from the service tank to the kiln through pipes, and then eight burners spray the fuel inside the kiln.

Slaking is carried out by hand on a platform. A concrete platform about 23cm above the ground has been constructed to avoid contamination from the soil. Quicklime is poured onto the platform and water is added using watering cans. If the lumps are too large then the layer of quicklime should not exceed 15cm.

After the quicklime has been slaked it is sieved, milled and then packed in 3-ply paper bags (50kg) ready for sale.

Marketing aspects

Lime produced by this project is sold for various uses which include sugar factories, soil stabilization, pulp, tanneries and the building industry. Most of the lime produced by the project is used by sugar factories as follows:

	Tonnes/year
Tanganyika Planting Company (TPC) Kilimanjaro	2500
Kagera Sugar Company (BUKOBA)	2200
Mtibwa Sugar Company (MOROGORO)	2150
Kilombero Sugar Company I & II (MOROGORO)	5000
Other users include:	
Mwanza Tanneries	1500
Tanganyika Farmers Association (TFA)	3000
Local consumers	N/S

The current price for one tonne of lime is TSh25 000 (US$108).

The project now produces half of the country's requirement for lime used in various applications. Some 350 workers are employed by these units.

Some of the problems facing the project are as follows:

- lack of modern machinery for hydration, sieving and packing
- power and water cuts
- lack of trained workforce
- lack of crusher and explosives to break the stones.

Management

The sponsor (SIDO) undertakes all managerial and supervisory duties of the project. In addition, a production manager competent in lime production, a project accountant, a storekeeper and 10 operators are employed.

This project, together with two other SIDO projects which use wood as fuel, have reduced the need to import lime, especially for industrial use and the building industry.

18. Small-scale lime processing: the Balaka experience

BRIAN JONES*

Balaka is a centre of small-scale limemaking in Malawi. There is a large limestone deposit at Chenkumbi Hills, some 15km from Balaka, which provides the raw material. The limestone is a poor, fractured deposit, but is good enough to produce reasonable lime.

During 1988, ITDG constructed a vertical shaft kiln in the Chenkumbi Hills.

---

*Aptech, Harare, Zimbabwe.

This used commercially produced softwood charcoal as fuel, and produced a higher quality of burnt lime with considerably improved fuel economy.

In 1989, Aptech was commissioned to design a plant which would be capable of hydrating and classifying 500kg/hr of quicklime from the kiln and producing 400kg/hr of lime hydrate at minus 150 microns. The plant was to be designed for minimum capital cost.

The plant was to produce two different grades of lime: high quality material with a high available lime content, around 60 per cent, which could be used in the sugar industry where it woud attract a price premium; and lower quality material which could be used for building or agriculture.

This chapter describes the design of the plant, its installation in Balaka in February 1990, preliminary commissioning in March and April 1990, and the subsequent history of the plant.

Description of plant

The plant at Balaka consists of three separate sections: the kiln, hydrator and the classification and milling plant. These are shown on the flowsheet and described below. The descriptions apply to buildings and equipment as designed and installed. Subsequent modifications are described under Installation and Commissioning below.

The function of the kiln is to calcine the limestone to form quicklime. It is described in Chapter 15 by John Spiropoulos.

Hydrator

The function of the hydrator is to hydrate the screened quicklime from the kiln.

Type	*Batch*
Length	2m
Trough diameter	800mm
Rotor shaft diameter	80mm
Liner thickness	6mm
Spoke pitch	80mm
Rotation speed	26.5rpm
Motor size	7.5kW
Quicklime feedrate	200kg/batch
Hydrated lime production	240kg/batch

Quicklime is loaded into the hydrator from the top in 200kg batches, and water added through holes in the bottom of water pipes running along the top of the machine. The rotor mixes the quicklime and the water. The hydrated lime is unloaded from the bottom.

Classification and milling plant

The function of this plant is to separate the incoming screened hydrated

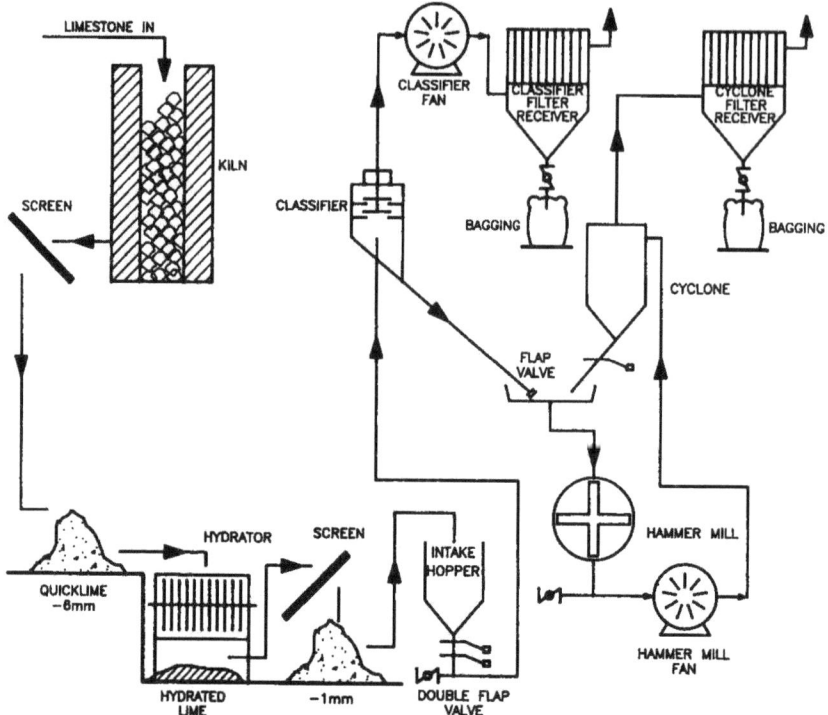

Figure 1: *Balaka kiln – flowsheet*

lime into plus and minus 150 micron fractions, and to mill the oversize to below 150 microns. The plant as designed consisted of an intake hopper, classifier, classifier fan, classifier filter, hammermill, cyclone and hammermill filter.

Lime is fed into the intake hopper, which discharges through a double flap valve into a pipeline which conveys it to a classifier. This separates it into coarse and fine fractions. The oversize material passes down a chute into the hammermill. The chute is fitted with a rubber flap valve which maintains the vacuum in the system but releases the coarse lime for milling.

The hammermill mills the oversize lime and blows it to the cyclone, which separates it into product size material and oversize. The oversize is recycled back to the mill, and the product size is blown to the filter.

The dusty air enters the filter tangentially at the top of the hopper and passes through the tubesheet to the inside of the bags. The clean air passes through the bag fabric to exhaust to atmosphere, leaving the dust to fall into the bagging hopper at the bottom. A butterfly damper was fitted to the nozzle to control flow of material during bagging.

The purpose of the enclosure was partly to protect the bags, and partly to give the option of rearranging the flow so that the dusty air was on the outside of the bags. It also enabled the filter to be run under vacuum if necessary, with the classifier fan downstream. The filters were fitted with a cleaning mechanism which consisted of a lever which enabled an operator to lift the bottom tubesheet momentarily and drop it onto the support ring. The shock of the drop was intended to shake dust from the inside of the bags.

The building was an open structure with a concrete slab floor, brick columns, gumpole purlins and a steel sheet roof. The classifier and filters were located on a 2.5m-high platform at one end of the building, with the hydrator and mill at ground level. The quicklime for the hydrator was stored on a platform at the other end.

Equipment specifications

Classifier
Type	Double whizzer
Diameter	450mm
Rotation speed	530–830rpm
Liner thickness	6mm
Motor size	0.75kw
Lime feedrate	400kg/hr

Classifier fan
Type	Centrifugal
Flowrate	1000m³/hr
Pressure rise	4.5kPa
Speed	3000rpm
Motor size	5.5kW

Hammermill
Type	Modified 1 tonne/hr maize mill
Model	4M, Series 24
Rotation speed	5500rpm
Motor size	22kW
Screen	304SS, with 2.5mm holes
Hammers	Tungsten carbide tipped
Lime feedrate	200kg/hr

Cyclone
Type	High efficiency Stairmand
Diameter	200mm
Length	800mm
Liner plate	6mm
Air flowrate	1000m³/hr

Filters
No. of bags	19
Total area	19m²
Face velocity	0.015m/sec
Air flowrate	1000m³/hr

Installation and commissioning

Hydrator

The installation and commissioning of the hydrator went well, except for problems with fumes and blockages in the stack. Most of the commissioning work went into finding the best operating procedure.

Water addition

Initially, water was added at a high rate (30 litres/min) for a few minutes, and the feed then tapered off. This created large clouds of dust after two minutes. The rate of water addition was reduced to 10 litres/minute. At this rate, the reaction still peaks at about 2 minutes, but the reaction is much less vigorous, and there is much less lime in the emission.

Initially, 100 litres of water were added for a 200kg batch. During the tests, it rapidly became apparent that less water was required, and the amount finally used was 55 litres per 200kg of quicklime.

Water pressure

Trials were carried out with water being fed from a 7m high storage tank and from a pump, to see if the water had to be at high pressure to keep the pipe holes open. No difference in hydration was observed.

Batch time

The initial hydration time was set at 20 minutes. As the trials progressed, however, this was reduced to 10–12 minutes.

Quality of product

The product is very homogeneous because of the mixing action of the hydrator. Available lime content varied from 50 to 60 per cent during the trials.

Hydrator speed

The speed of the hydrator is presently 26.5rpm. The batch size could be increased by 50 per cent by reducing the speed of the hydrator to about 18rpm.

Pollution

During the early runs of the hydrator, the 100nb vent stack appeared to work well, but started to block as steam condensed on the walls and collected lime dust. The operators tried a number of modifications, but without much success. The hydrator needs a permanent, large-bore vertical vent with cleaning access.

The hydrator lid and bottom door did not seal well, and emitted steam and lime dust during hydration. These should be redesigned to improve the seal and increase rigidity.

Operability

The hydrator crew consisted of two shifts of four operators and a supervisor. The hydrator and classification sections worked only on days, since this was adequate to process the production from the kiln.

Classification system

The intake hopper and valve fed material into the system, but frequent blockages took place in the tee-piece and the long radius bend leading to the classifier. It was eventually found that this was caused by the classifier fan not performing to specification. Despite this, the operating crew gained a lot of experience and, by the end of the week to 15 March, could reliably feed lime in at 450kg/hr.

The classifier was initially run at 800rpm, but screen analyses indicated that the classifier was making a very fine cut (99.9 per cent less than 63 microns) at this speed. The speed was reduced to 670rpm and then to 550rpm, with corresponding increases in classifier fine particle size and the percentage of the feed going to the classifier filter. This was not sufficient so, on 12 March, one of the whizzers was removed. Later, the diameter of the remaining whizzer was reduced by 50mm. Finally the system was run on 13 March with the classifier whizzer stationary. The results from this test showed a reasonable 150 micron cut, with 5 per cent oversize in the classifier filter and about 12 per cent undersize in the hammermill feed.

This unusual result was also explained by the poor performance of the classifier fan. A characteristic curve test carried out on 15 March showed that the flowrate and pressure rise in the fan were significantly below both the specification and the fan curve. The fan was actually a clean air fan, where a dirty air fan had been specified, and the narrow passages of the closed impeller had clogged with lime. The fan also suffered from vibration.

Numerous problems were experienced with the hammermill, including falling hammers, missing bolts and worn bearings. Initially the most serious problem was the slow feedrate into the mill. To combat this, the angle of the feedtray was increased to 10° and the feed hopper was turned around on the feed tray so the lime had a shorter and more direct route into the mill. The feed of material into the mill, and the operation of the mill are both now good.

Gaskets were fitted on all of the joints, and a seal fitted to the shaft to reduce dust emission.

The hammermill cyclone was operated during the early commissioning runs, but was soon disconnected because of the heavy emissions of dust from the bottom of the cyclone through the flap valve.

The initial trials of the filter went very poorly, with extensive leakage from both filters. This was primarily a fabrication problem. The filter tubesheets have now been gasketed and bolted to the support ring. This works well. The filter is manually cleaned.

It was soon found that lime in both hoppers hung up and would not discharge easily, even when the hopper was hit with pieces of wood. The lime is not sticky, but seems to have an angle of repose which can exceed 90°.

A problem of dust discharge during bagging was overcome when the operators fitted a canvas sheath around the bottom of the discharge nozzle, going down into the bag. This is a traditional dust control measure on maize mills in Malawi, and reduced dust on the plant quite significantly.

Production trials

Following initial production trials in April 1990, further trials were conducted at the beginning of June 1990, when it was found that the operators had changed the procedure completely, as detailed below:

Calcine limestone in kiln
Hydrate all of the quicklime discharged from the kiln
Screen hydrated lime to remove +6mm material
Classify and mill −6mm lime
Dump +6mm lime

This change of operation increases the output of the plant considerably, with the ratio of bagged product to kiln feed rising from 35 to 68 per cent. It also involves running most plant items, like the hydrator and the mill, with feed material sizes considerably over design, but this does not seem to cause any serious problems.

Screen analysis of the −6mm fraction indicated that 70 per cent was already product size (minus 150 micron). This makes classification of this fraction to remove the fines worthwhile.

The improved material balance is shown in Figure 2.

The operators made a number of changes to the plant in attempts to overcome operational problems. These are summarized below:

1. A rotary feeder was fitted in place of the double flap feeder. This improved the feed of lime into the classifier.
2. The bagging hoppers were replaced by trouserleg hoppers with a hopper angle of 75°. This solved the problem of lime hangup.
3. The classifier fan was fitted on the downstream side of the filter, so that it could act as a clean air fan.
4. The loading hopper was relocated up on the filter platform, 2.5m above ground level, next to the classifier to make the transport of lime into the classifier easier. This makes the loading of the hopper much more difficult.
5. The classifier was replaced with a vertical axis vane classifier. After considerable modifications this started to operate satisfactorily. The old

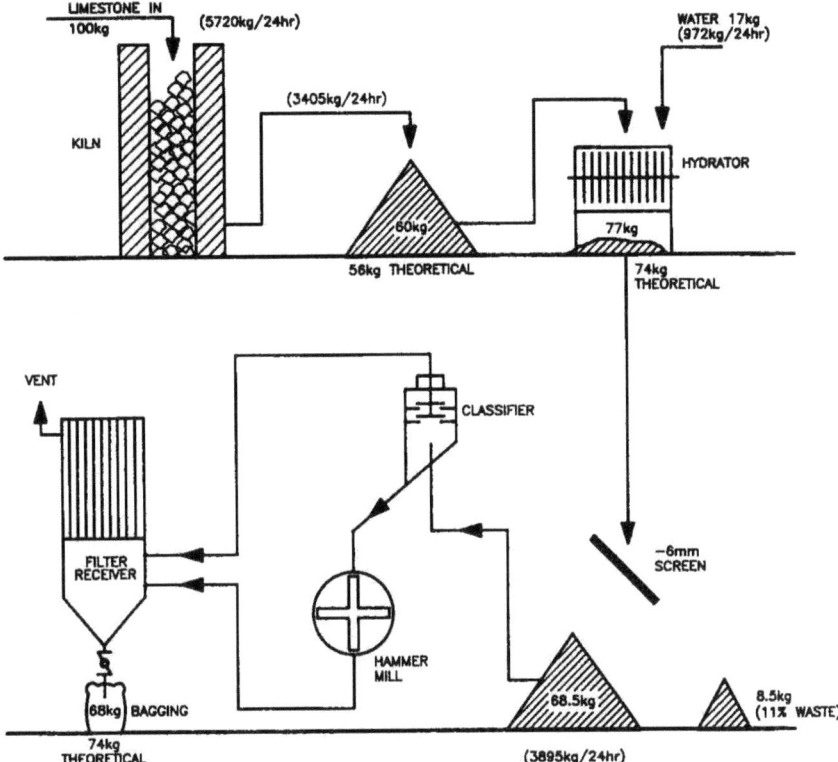

Figure 2: *Balaka kiln – material flow*

classifier was installed in the line from the hammermill to the filter to recycle oversize, with a second fan installed on the clean side of the filter to draw lime through to the filter.

Most of these actions were taken, at considerable cost, to overcome problems caused by the poor performance of the classifier fan.

Design implications for a future plant in Malawi

Kiln

The kiln works well. The steel casing is not necessary, and makes the use of the access ladder uncomfortably hot when the kiln is firing. A future design would use steel straps to reinforce the brickwork. Similarly, the reinforced concrete ring beam is not necessary.

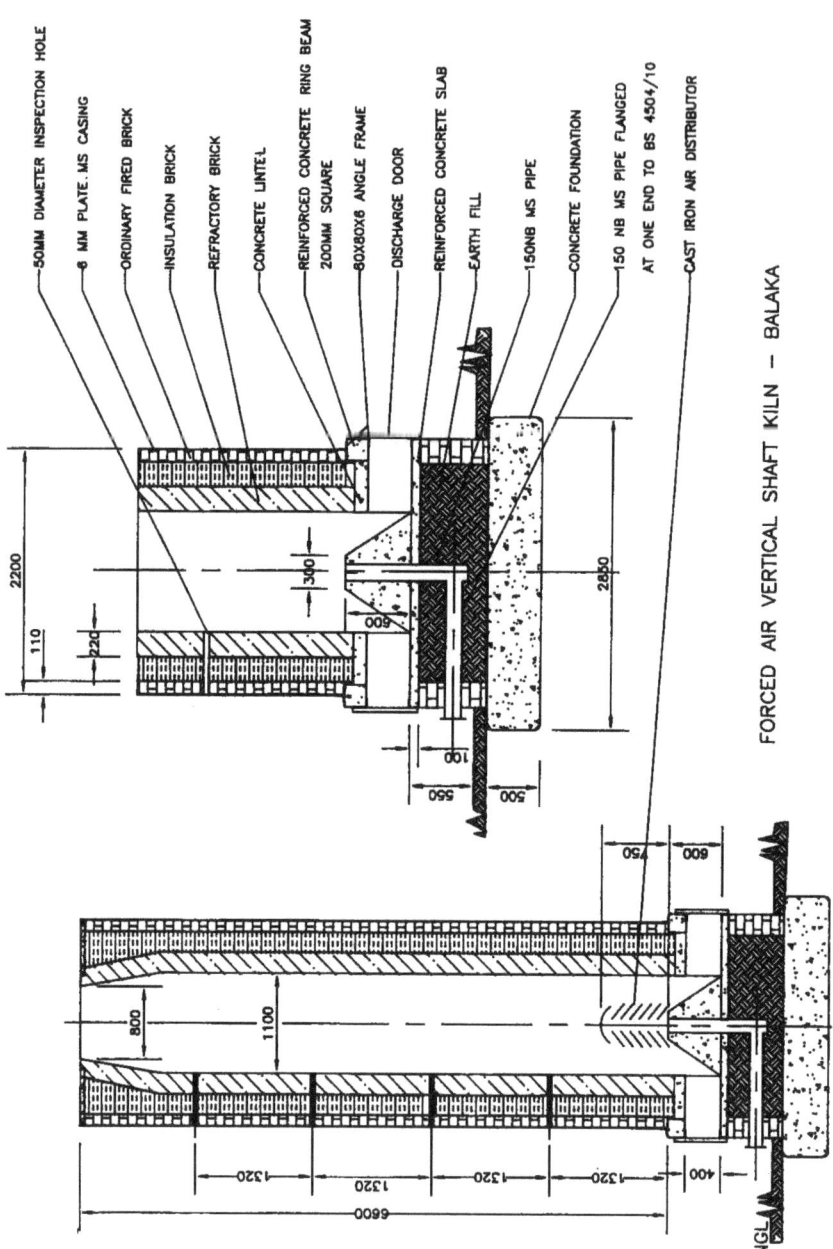

Figure 3: *Balaka forced air vertical shaft kiln*

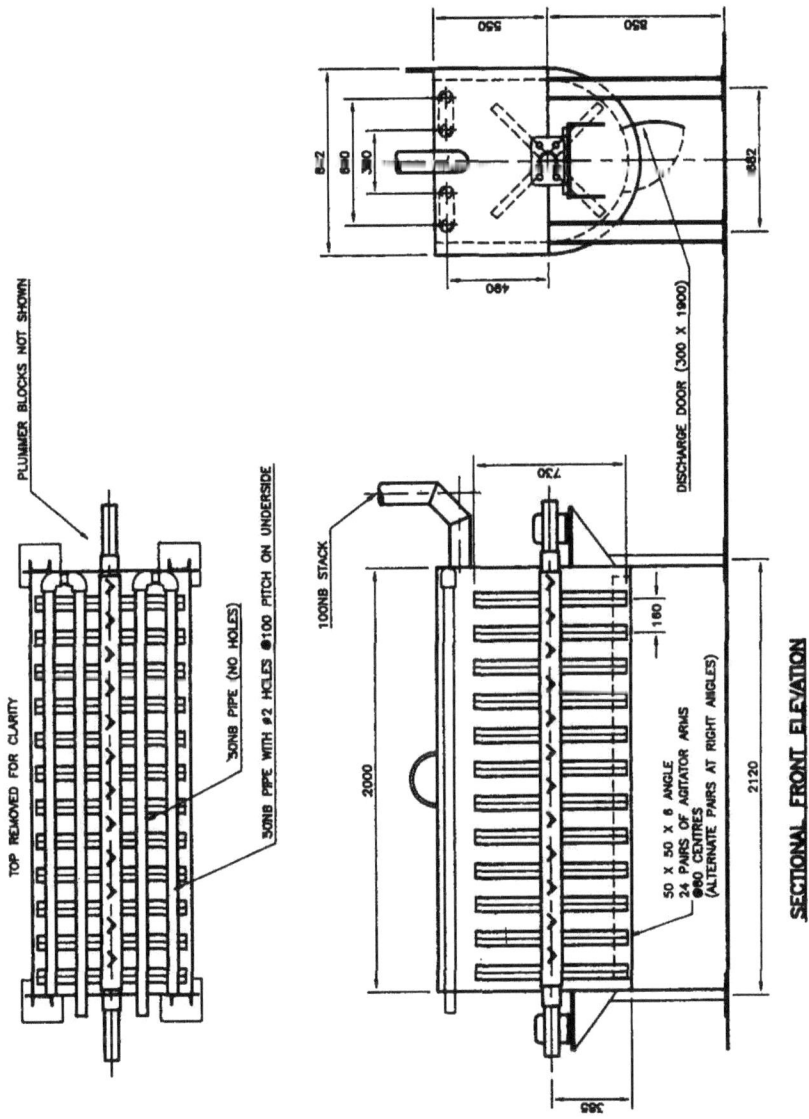

Figure 4: *Lime hydrator layout*

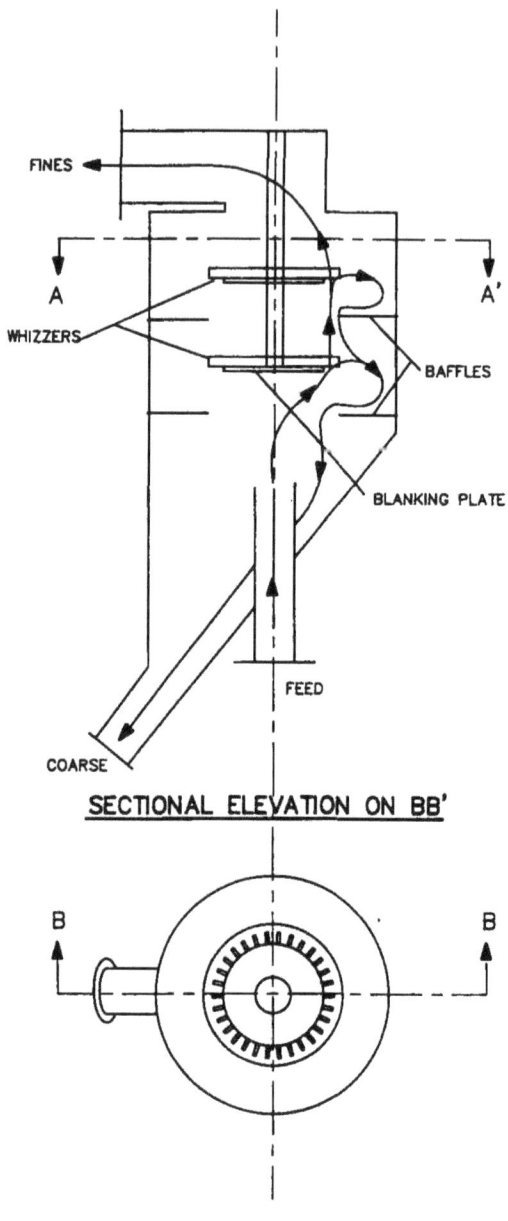

Figure 5: *Twin whizzer classifier*

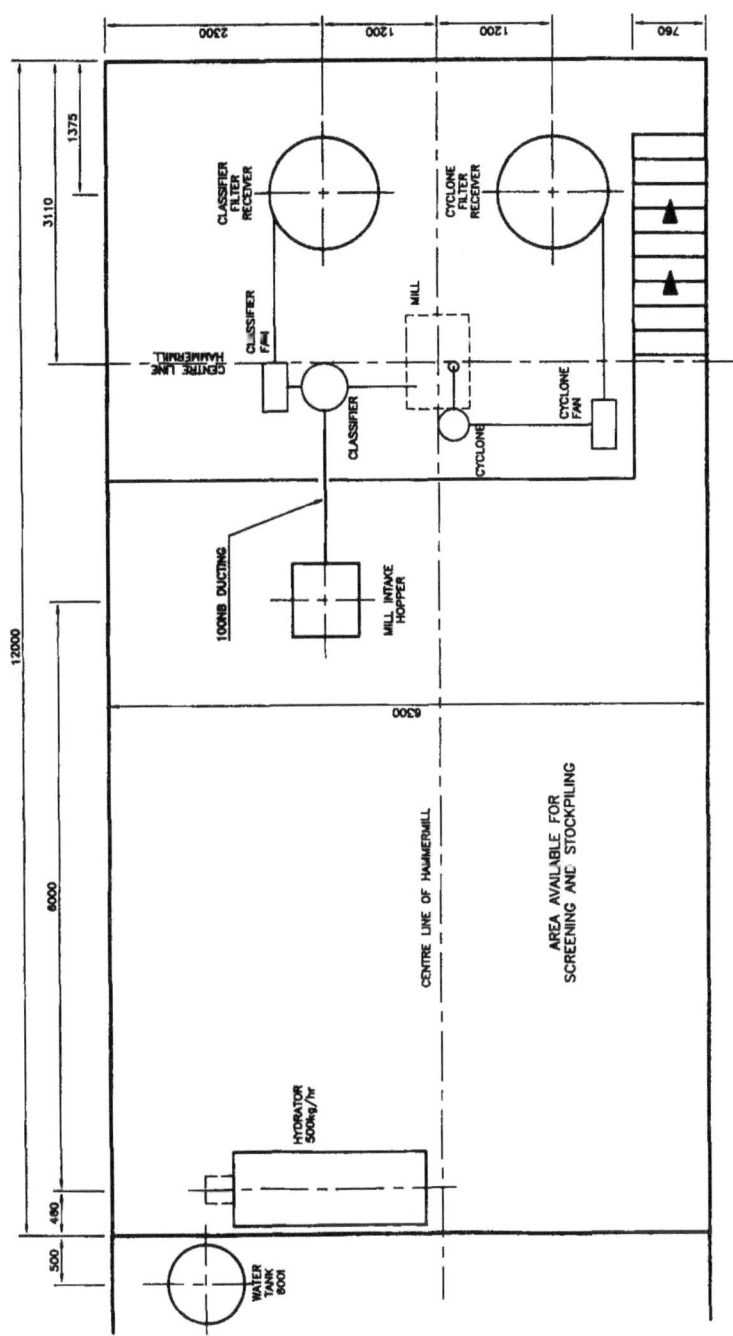

Figure 6: *Balaka plant layout plan*

Hydrator

The hydrator was specifically designed as a batch machine in order to test the principle of hydrating in this manner. A future machine should be designed as a continuous flow unit to increase production rates and reduce venting problems.

The trough liner is not necessary, and should not be used in a future design.

Classification and milling system

Intake system

Commissioning trials established that the original intake worked at the design rate, in a situation where the classifier fan was not performing to specification. Nevertheless, it was a delicate operation, and it was worthwhile installing the rotary valve in the place of the double flap valve to feed material into the system. This is a more reliable feed mechanism.

Classifier

Both the original and the new classifiers are capable of operating to specification, and are commonly used in lime plants. The whizzer classifier is better in a situation where core particles of limestone are surrounded by hydrated lime, because the impact with the blades tends to shake off the hydrate, making separation easier. The vane classifier makes a slightly more accurate size separation.

Classifier fan

The operation of the fan is crucial to the good operation of the classification system. If an adequate dirty air fan is not available, the best place for the fan is on the clean side of the filter. This needs to be conservatively rated to ensure reasonable performance.

Filter

Both filters now work well. Increasing the angle of the bagging hoppers to 75° has resolved the problem of lime hangup.

Hammermill

With the hammermill operating at the design rate, the duty of the hammers is quite arduous, and high wear rates should be expected. The choice of material for the hammers varies from the inexpensive standard hammers which will wear quickly to hammers made from more exotic materials which will give longer life at higher cost. Once a variety of materials has been tried, a decision should be taken on the basis of the lowest cost per tonne of milled product.

Gaskets are necessary on all joints on the mill, as well as a seal on the rotor shaft where it enters the hammermill fan casing, in order to minimize dust emission. Sealed bearings are necessary on the rotor shaft.

With a high proportion of fines, the mill fan impeller blades and casing liner plates wore rapidly. The removal of the impeller and replacement by a fan on the clean side of the filter removes this problem.

The feed hopper modifications made on site are essential to ensure the smooth flow of material into the mill.

Hammermill cyclone

This cyclone was installed in an attempt to recycle oversize material from the hammermill without using a separate fan, in order to reduce costs. The replacement by a second classifier and fan gives much better performance.

Conclusions and recommendations

The plant operates satisfactorily, and is in use. The owners plan to expand production. The experience of the Balaka plant shows that it is possible to design, build and operate a small-scale plant producing high quality lime to a controllable size specification in a country like Malawi. It can operate safely and with high fuel efficiency.

Plant design

The design of a future plant should be based on the Balaka kiln and classification system, with a continuous flow hydrator based on the batch hydrator. Equipment specifications should be based on the operating procedure introduced by the operators at Balaka.

The design of the plant is easily modified, so the capacity of the plant should be based on the market for lime in the area to be supplied by the plant, and the capital resources of the owner.

The configuration of the plant should be based on the market for lime, the prices paid for different grades of lime, and the proportion of product size lime in the feed to the classifier. The design should aim to give the optimum value product mix at the lowest capital cost. It should also be flexible enough to allow operators to change the product size easily to meet changing market conditions.

Fabrication and installation experience

A number of problems were experienced in the fabrication and installation of equipment for the plant. These problems are not unique to Malawi, and the conclusions drawn from them may be useful in future projects.

Engineering companies often have good equipment and material stocks, but artisan skills and workshop supervision can be weak. This produces defective work in both fabrication and installation. When these problems occurred in this project, the fabricators admitted liability immediately and rectified quickly, often at considerable cost to themselves. There is no doubting their commitment to the project. But a lot of the work would not have been necessary if workshop inspection had been better.

In order to reduce these problems, engineering drawings should be very comprehensive, and not assume detailed knowledge on the part of the artisan. Inspection in the factory and supervision on site should be more frequent to reduce defective work.

Operations

The commissioning of the plant should be a joint effort by the plant designers and the operators. This will significantly reduce operating problems after commissioning.

Considerable effort must be put during commissioning into training operators and supervisors in safe and proper working procedures until good working habits have been established.

Operation of both kiln and plant is within the capacity of local people, given adequate training. The workers on the plant were keen and helpful, and this made setting up production much easier, but the plant requires people to work in ways which are quite different to customary work patterns. The plant requires shift working, and all parts of the plant need work to be carried out in a systematic, steady way to reach a reliable production rate. The consequence of error, especially on the kiln, can be quite serious. This was demonstrated when an error in loading coal caused a burnout of the kiln lining in June 1990.

The modified operating procedure introduced by the operators significantly increases productivity and should be used in future plants.

19. Portland-pozzolana cement from sugarcane bagasse ash

SYED FAIZ AHMAD* and ZAHEERUDDIN SHAIKH*

Sugarcane bagasse ash

The family of neo-pozzolanas includes rice husk ash (RHA), rice straw ash (RSA), wheat straw ash (WSA) and others. Sugarcane bagasse ash (SCBA) can be classified as a new entrant to the clan of neo-pozzolanas.

---

*National Building Research Institute, Karachi, Pakistan.

Two samples of SCBA were collected from the boiler houses of two different sugar mills in the province of Sindh and were marked as Sample-M and Sample-F respectively. The experimental programme embodied the following:

- assessment of the quality of SCBA
- properties of the SCBA mortar both fresh and hardened
- resistance of SCBA to sulphate attack, and
- pozzolanic activity index.

Quality of SCBA

The test results of the physical and chemical analysis are shown in Tables 1 and 2.

Chemical properties

Table 2 shows that the cumulative percentage of major oxide contents is high in both cases of Sample-M and Sample-F of SCBA. These values are higher than the minimum value of 70 per cent laid down by the ASTM standard on cement 618–78 for class N pozzolanas. The sulphate content, as SO_3, is quite low and it is 1.01 per cent and 0.28 per cent for Samples M & F respectively. This is much lower than the maximum value of 5 per cent laid down for class N pozzolanas in ASTM 618–78.

The magnesium oxide (MgO) is determined to be 2.4 and 3.03 per cent

Table 1. Chemical composition of sugarcane bagasse ash (SCBA) Samples M & F and comparison with OPC & RHA

Composition		SCBA (%) Sample-M	SCBA (%) Sample-F	RHA[2] (%)	OPC[2] (%)
Silicon dioxide	(SiO_2)	60.28	63.89	92.05	20.90
Aluminum oxide	(Al_2O_3)	9.29	8.95	0.94	5.20
Iron oxide	(Fe_2O_3)	1.72	3.49	0.81	4.10
Calcium oxide	(CaO)	7.98	7.21	0.27	63.70
Magnesium oxide	(MgO)	2.40	3.03	0.27	1.80
Sodium oxide	(Na_2O)	2.84	1.46	0.06	0.14
Potassium oxide	(K_2O)	8.15	8.34	1.72	0.71
Sulphate	(SO_3)	1.01	0.28	0.31	2.30
Loss on ignition		5.11	2.52	3.19	1.00
Specific gravity		2.37	2.44	2.09	2.09
Fineness:					
– Wet sieving, finer than sieve No. 325		88	96	93	84
– Blaine, cm²/g		–	–	14 300	3000

Table 2. Physical and chemical properties of sugarcane bagasse ash (SCBA) Samples M & F and others in relation with the requirements of ASTM C 618–78

Property	ASTM C 618–78 classes of pozzolana			SCBA Sample-M	SCBA Sample-F
	(N)	(F)	(C)		
Chemical properties					
$SiO_2 + Al_2O_3 + Fe_2O_3$	70.0	70.0	50.0	71.29	76.33
SO_3 (max %)	4.0	5.0	5.0	1.01	0.28
MgO (max %)	5.0	5.0	5.0	2.40	3.03
Na_2O (max %)	1.5	1.5	1.5	2.84	1.46
Loss on ignition	10.0	2.0	6.0	5.11	2.52
Physical properties					
Moisture content (%)	3.0	3.0	3.0	0.86	0.58
Fineness % retained on sieve No. 325	34.0	34.0	34.0	4.00	12.00
Pozzolanic activity index with OPC at 28 days (% of Control)	75	75	75	84	84
Pozzolanic activity index with lime at 7 days, psi (MPa)	800 (5.5)	800 (5.5)	800 (5.5)	650 (4.5)	650 (4.5)

respectively and it also meets the requirements of ASTM class N pozzolanas, which stipulates a maximum value of 5 per cent.

The alkali content (Na_2O) in case of Sample-M is shown to be a little higher (2.48 per cent) while that of Sample-F is low (1.46 per cent) and is well within the maximum value of 1.50 per cent as laid down for ASTM class N pozzolanas.

The loss on ignition is indicative of total carbon content. The value obtained in cases of both Samples M & F are shown to be very low, 5.11 per cent and 2.52 per cent respectively as against maximum value of 10 per cent laid down for ASTM class N pozzolanas.

Physical properties

The value of specific gravity obtained for Samples M & F are shown to be 2.37 and 2.44 respectively.

The SCBA Samples M & F were both ground in a ball mill. After wet sieving as per ASTM C430–78, the resulting fineness was found to be 88 and 96 per cent finer than sieve No. 325, for Samples M & F respectively. The fineness of ordinary Portland cement used, also determined through wet sieving, was shown to be 84 per cent finer than No. 325 sieve. The finenesses of the SCBA used were quite high and were greater than the requirements of ASTM 618–78 for class N pozzolanas, which lays a minimum value of 66 per cent finer than sieve No. 325.

The normal consistency of pastes containing various proportions of SCBA with OPC was determined as per ASTM C 187–79. For comparison the normal consistency of neat OPC paste was also determined. The results are shown in Table 3.

It can be seen that the value of normal consistency of the pastes increases as the proportion of SCBA in the paste increases. This means that with the addition of more and more SCBA, an increased amount of water is required to obtain the desired consistency. It is in complete conformity with the normal behaviour of the pozzolanas.

Table 3. Values of normal consistency of pastes of neat OPC and pastes containing OPC with SCBA in varying proportions

Proportion OPC:SCBA	Normal consistency %
100:00	27.0
90:10	28.5
85:15	29.0
80:20	29.0
75:25	29.5
70:30	29.5

Strength of SCBA pozzolanas

The strength characteristics of hardened SCBA mortar have been determined as per ASTM C 109–77. The proportioning of SCBA with OPC and lime has been done as per schedule as shown in Tables 4 and 5.

Substitution of OPC with SCBA

Blended cements were produced by interblending OPC and SCBA partially replacing cement at a rate of 10, 15, 20, 25, and 30 per cent by weight. Both Samples M & F of SCBA were used. The schedule of proportion is shown in Table 4.

The compressive strengths at 10 different ages, namely 1, 3, 7, 14, 28, 56, 90, 140, 224 and 365 days, were determined as per ASTM C 109, using 50mm mortar cubes and the water-cement ratio having been governed by normal consistency values determined for different SCBA substitution cases.

For SCBA Sample-M the test results of the two batches (MA & MB) representing 10 and 20 per cent replacement of OPC become equal to that of the control batch (C) at the age of 224 days. And at the age of 365 days these become almost equal to one another. On the other hand, at the age of 90 days all batches of Sample-M, i.e. even at 30 per cent replacement of OPC, surpass the requirements of the blended cements types S, IS, P, and IP of ASTM C 595, as shown in Table 6.

Table 4. Schedule of batches using OPC and SCBA

Batch title	Proportion OPC:SCBA % by weight	No. of specimens tested at each age (Nos.)	Ages of specimens tested (days)	Total specimens tested (Nos.)
	I. Sample-M			
MA	90:10	03	1, 3, 7, 14, 28,	30
MB	85:15	03	28, 56, 90, 140,	30
MC	80:20	03	224 and 365	30
MD	75:25	03	(10 different	30
ME	70:30	03	ages)	30
	II. Sample-F			
FA	90:10	03		30
FB	85:15	03		30
FC	80:20	03	"	30
FD	75.25	03		30
FE	70:30	03		30
	III. Control batch			
C	100:00	03	"	30

Table 5. Schedule of batches using OPC with SCBA and lime

Batch title	Proportion OPC:SCBA:LIME % by weight	No. of specimens tested at each age (Nos.)	Ages of specimens tested (days)	Total specimens tested (Nos.)
	I. Sample-M			
MF	90: 5: 5	03	1, 3, 7, 14, 28,	30
MG	80:10:10	03	28, 56, 90, 140,	30
MH	70:15:15	03	224 and 365	30
MI	60:20:20	03	(10 different	30
MJ	50:25:25	03	ages)	30
	II. Sample-F			
FF	90: 5: 5	03		30
FG	80:10:10	03		30
FH	70:15:15	03	"	30
FI	60:20:20	03		30
FJ	50:25:25	03		30

In case of SCBA Sample-F, the results are even better. At the age of 224 days, the test results of the two batches (FA & FB) representing 10 and 20 per cent replacement of OPC become almost equal to that of the control batch (C). And some even surpass the value of the control at the age of 1 year. Moreover, a better overall performance is observed in case of all other batches of SCBA Sample-F. In this case at the age of 28 days the test results of all the batches, i.e. even at 30 per cent replacement of OPC, either become equal to or surpass the requirements of the blended cements type S, IS, P, and IP of ASTM C 595. And at the age of one year, the values of all the batches of Sample-F become almost equal to the value of the control batch.

A better performance for SCBA Sample-F is logical because the SCBA Sample-F was ground to a greater fineness. This confirms the philosophy that the pozzolanas are more reactive if they are blended with cements in a very finely ground state.

Substitution of OPC with SCBA and lime

In this case, blended cements were produced by interblending OPC with SCBA and lime. SCBA and lime were blended in a ratio of 1:1 and these two were then used to replace OPC at the rates of 10, 20, 30, 40 and 50 per cent by weight. The schedule of proportion is shown in Table 5. Both Samples M & F of SCBA were used for the purpose.

In this case also, the compressive strengths at 10 different ages were determined using 50mm mortar cubes.

For SCBA Sample-M, at 30 per cent replacements of OPC, the requirements of the ASTM type S, IS, P, IP are met at an earlier age of 56 days only as against 90 days for SCBA–OPC mortars. And by 90 days the strengths at 40 per cent replacement equal the requirements of the ASTM type S, IS, P, IP. The results at 50 per cent replacements, however, are not satisfactory; this may be attributed to inferior fineness of SCBA Sample-M.

The SCBA Sample-F once again gives a better performance. At the age of 28 days, the strengths at 30 per cent replacement meet the requirements of ASTM type S, IS, P, IP. And more appreciably, at the age of 56 days, strengths at 50 per cent replacement of OPC surpass the requirements of ASTM type S, IS, P, IP. And, at the age of one year, the strengths up to 40 per cent replacement of OPC almost equal the strength of the control batch.

Table 6. Compressive strength requirements of blended cements

Type of blended cement	Compressive strength requirements as per ASTM C 595, PSI (MPA)		
	3 days	7 days	28 days
Type S	–	600 (4.1)	1500 (10.3)
Type IS	1800 (12.4)	2800 (19.3)	3500 (24.1)
Type P	–	1500 (10.3)	3000 (20.7)
Type IP	1800 (12.4)	2800 (19.3)	3500 (24.1)

Resistance to sulphate attack

There are no standard methods of test for sulphate resistance of blended cements. In this case, the experimental study was designed to examine the ability of SCBA-mortar to resist sulphate attack. Two corrosive media, one having a 5 per cent magnesium sulphate solution and the other a 10 per cent magnesium sulphate solution, were prepared in five different containers. In these containers, different specimens were immersed for long-term study. From Table 7, it can be seen that the specimens dipped included those of SCBA Samples M & F with curing ages varying from 103 days to as low as 11 days. The idea was to study the effect of varying proportions of SCBA and the age of mortar in resisting the sulphate attack. For comparison, specimens of the control batch were also immersed. The visual observation on deterioration were recorded for one year. From the results of examination, it appeared that the mortars containing SCBA perform better than the mortars containing OPC alone.

Pozzolanic activity of SCBA

The Pozzolanic Activity Index (PAI) with ordinary Portland cement (OPC) was determined as per ASTM C 311-77, and it is defined as the ratio of the compressive strength at 28 days of the test specimen to that of the control specimen, expressed in percentage. The test result indicated that the PAI was 84 per cent which is more than the minimum requirements for ASTM class N pozzolana; the minimum value stipulated is 75 per cent.

Table 7. Sulphate attack test, test specimen details

(A) 5% Magnesium sulphate solution			(B) 10% Magnesium sulphate solution		
Cube no.	Proportion of SCBA (%)	Age at dipping (days)	Cube no.	Proportion of SCBA (%)	Age at dipping (days)
C 16	Control	103	C	Control	35
MA 7	10	98	FA	10	21
MB 17	15	97	C	Control	15
MC 16	20	96	FB	15	14
MD 8	25	85	FC	20	11
ME 16	30	84			
FA 15	10	82			
FB 16	15	78			
FC 4	20	69			
FD 19	25	64			
FE 4	30	55			
C 6	Control	53			
MA 18	10	51			
FA 9	10	49			

The PAI with lime is defined as the compressive strength at 7 days of the test specimen expressed in psi (MPA). This property was also determined as per ASTM C 311–77 and the value obtained was 650psi (4.50 MPa). This is a little lower than the minimum value specified for class N pozzolanas; the value specified is 800psi (5.5 MPa). It may be added here, however, that, according to ASTM, neither the PAI with OPC nor the PAI with lime is to be considered as a measure of the quality of pozzolana and hence the compressive strengths of concrete or mortar.

Conclusions

The following conclusions are drawn from the results of this investigation on the blended cements having OPC + SCBA and OPC + SCBA + lime at various percentage replacements:

1. The strength development characteristics of the SCBA pozzolana indicate that, during first 14 days of hydration, pozzolanic reactions did not make much contribution to the strengths. However, the evidence of pozzolanic activity is more pronounced at later ages, especially during 28–365 days, when cement containing 30 per cent SCBA also became almost equal to the control.
2. In the case of OPC-SCBA-lime blend, the performance is even better, the blend having 40 per cent replacement giving results almost equal to the control.
3. The physical and chemical properties of SCBA are satisfactory and conform to the requirements of ASTM class N pozzolanas.
4. The visual observations on deterioration of the mortar cubes containing SCBA at various proportions and curing ages when immersed in the magnesium sulphate solution for a prolonged duration of one year showed little or no effects on samples containing an increased proportion of SCBA. On the other hand, samples of control batches showed marked deterioration. This establishes the fact that SCBA provides resistance to sulphate attack.
5. The Pozzolanic Activity Index (PAI) with OPC was determined to be 84 per cent which is more than what is laid down for ASTM class N pozzolana. However, the PAI with lime showed a little lower value as compared to the stipulations of the ASTM Standard.

20. KVIC technology in the production of lime and alternative cements in India

P H NAIK*

The task of the development of village industries was assigned to the Khadi and Village Industries Commission (KVIC), a Government of India organization set up under the Khadi and Village Industries Act, 1956, which came into being on 1 April 1957. The programme of KVIC extension services is directed at intensive efforts for rural industrialization. The aims of KVIC include creating self-employment opportunities, producing saleable goods and creating self-reliance among rural people (particularly artisans) and building a strong rural community spirit. The broad objectives are generating employment opportunities in rural areas, skill development through training programmes and transfer of technology for the purpose of rural industrialization.

The functions of KVIC include planning and organizing entrepreneurship development programmes, procurement and distribution of raw materials, machinery supply and marketing facilities. KVIC also promotes research and development in production techniques and transfer of technology. The basic function of KVIC is to promote the concept of co-operation.

The role of KVIC in the promotion of lime industry

KVIC has laid emphasis on the introduction of economic and improved lime kilns and their construction on scientific principles, supply of improved tools and equipment, training lime artisans in the use of improved techniques and the efficient operation of lime kilns, extension of concessional finance for the construction of improved lime kilns, procurement of tools and provision for working capital, extension of technical guidance for construction and commissioning of improved lime kilns and encouragement to co-operative organizations of the lime artisans.

The lime products in KVIC include hydrated lime, quicklime (CaO – chemical grade), building-grade lime, lime mortar, lime pozzolana (LYMPO – $CaSiO_2$) and chalk lime.

Improved lime kilns

After undertaking a thorough study of traditional lime kilns which were inefficient thermally as well as for calcination, KVIC evolved an improved lime kiln of one to 10 tonnes per day (t/d) output. For the benefit of lime artisans, designs of lime kilns of one t/d capacity for burning limestone has

---

*Khadi and Village Industries Commission, Bombay.

been standardized. Lime kilns designs in the capacity ranges of 1, 2, 3, 5 and 10 t/d have also been evolved by KVIC for individual entrepreneurs as well as co-operative societies. An appreciable number of KVIC designed lime kilns are operating in India as well as in Botswana, Bolivia, China, Philippines, Mozambique, Nigeria, Sudan, Tanzania, and Zambia.

KVIC lime technology

Technical aspects

For the efficient performance of a lime kiln in terms of thermal efficiency, optimum fuel consumption and full calcination, an efficient proportion between the effective height and the diameter has been maintained. This also facilitates the maintenance of the three zones, viz. pre-heating, calcination and cooling.

The inner shaft of the kiln has been lined with fire bricks to a width of 23cm for resisting high temperature, abrasion, corrosion and chemical erosion to extend kiln life.

The outer shaft is of thick, red brick masonry to minimize heat losses leading to lower fuel consumption and maximum heat retention. Insulation of loose sawdust, paddy husk or fired clay has been placed between the refractory lining and the outer wall to enhance thermal insulation.

Four discharging doors of specific size have been set at the bottom of the shaft for air inflow and to facilitate the discharge of quicklime.

A detachable G.I. chimney (in the case of the one and three t/d kilns) is provided at the top of the inner shaft to induce natural draught. All the four discharging doors have been provided with adjustable dampers to regulate the inflow of air, enabling proper draught control according to variable weather conditions.

An air injector which can be coupled to an electric air blower blows air into the charge for increasing the draught in kilns burning sea-shells.

Poke holes have been set into the wall on opposite sides for observing the thermal conditions and to regulate the process of calcination accordingly.

Mild steel bands (rings) are fitted around the outer periphery of the kiln. After allowing for the initial thermal expansion of the kiln, these bands are tightened to protect the kiln from thermal cracks.

A suitable elevating device, preferably an electric hoist, for speedy charging of the kiln and a suitable staircase for access to the top of the kiln are recommended.

Constructional aspects

Great care has to be taken at each stage of construction of the kiln, since the successful operation and efficiency of it depends greatly on the quality of construction.

A sound foundation consistent with the nature of the soil, the size of the kiln and the total weight of the superstructure of the kiln needs to be laid, employing boulders, ballasts, cement, etc. The superstructure consisting of the basement and the shaft should be built only after the proper setting of the foundation.

A brick or rubble masonry basement consisting of the plinth is laid over the foundation. In the case of kilns burning shells, the galvanized iron pipe leading to the air injector should be laid while the basement is made.

Four discharging doors of specific size are installed immediately after the basement. These doors are opposite each other and spaced symmetrically. Segmental arches above these doors are recommended.

The construction of the superstructure consisting of the inner refractory lining, insulation gap and the outer brick masonry shaft needs to be commenced simultaneously with the discharging doors.

The inner surface of the lining has to be smooth, and the joints of the bricks thin to eliminate heat seepage.

Provision for poke holes needs to be made at different heights of the superstructure. After completion of the superstructure, a sloping floor should be laid at the bottom of the shaft. This will facilitate discharge of quicklime.

Since the construction of lime kilns involves the laying of refractory as well as red bricks in circular form, it is advisable to employ highly skilled masons.

Operational aspects

The efficient and economic running of a lime kiln depends mainly on the skill of the kiln operator. The following are the hints for successful operation:

- Optimum fuel consumption by maximum utilization of heat energy generated by it.
- Elimination of heat losses by radiation, conduction and convection through the kiln walls, exhaust gases and quicklime discharged from the kiln. This can be achieved by maintaining the imaginary pre-heating, calcination and cooling zones at their appropriate levels. By the efficient control of the draught, these zones can be maintained. Only cool quicklime needs be drawn out of the kiln. Efficient draught control also envisages transfer of heat emitted by the hot quicklime at the cooling zone at the bottom of the shaft to supplement the heat energy requirement in the calcination zone in the middle of the shaft. The heat discharged out of the calcination zone is utilized in the pre-heating zone at the top portion of the shaft to pre-heat the charge of limestone or shells and fuel, thereby minimizing the heat requirement in the calcination zone.

- Charging of limestone and fuel of uniform size and the maintenance of a perfect proportion of limestone or shells, and fuel.
- Avoidance of a quick rise in temperature helps in preventing cracks and consequent heat loss and also prolongs the life of the kiln.

The KVIC-designed lime kilns offer fuel saving of 30–40 per cent over traditional lime kilns.

Quality control

Quality control is maintained as standard and regulated through laboratory tests. Manufacturing standards are set at the start for the continuous process. All standard norms, as per technical designs, are followed at the time of construction. KVIC has a laboratory for the testing of lime products produced by the implementing agencies in the country and their samples are being tested frequently and results communicated. Details of defects, if any, in the product are passed on to the entrepreneurs and guidance is given in maintaining quality control.

Pollution

KVIC kilns are mostly situated in rural areas and in the outskirts of villages and do not cause excessive pollution. However, different R&D laboratories have developed systems of pollution abatement to meet emission standards laid down by the Pollution Control Board.

Calcination of limestone requires temperatures exceeding 900°C and the required heat energy is usually supplied by firing coal along with the limestone. There are heaped and rectangular trenched kilns known as country kilns and the vertical shaft kilns of KVIC. The coal to limestone ratio varies from 17 to 30 per cent.

The kilns produce SO_2, CO_2 and some unburnt coal and limestone and coal dust emitting from the top of the kiln during charging. However, the KVIC kilns have batch feeding only twice a day and as such pollution is minimized.

Fuel

Steam coal is being used as fuel. Soft coke can also be used but the cost of production will be a little higher due to the higher price of coke. Further, KVIC kilns can also be run with firewood where coal is either scarce or firewood is abundantly available at a lower price. KVIC kilns have the advantages of optimum fuel consumption, full calcination of limestone, simplicity in erection, continuous operation, and non-polluting exhaust gas from the kilns all resulting in cost effectiveness.

KVIC has carried out experiments using firewood as fuel. The results obtained were economically encouraging. This fuel can be burnt efficiently in the varying KVIC kilns. Many KVIC kilns are operating on firewood all over India, and KVIC has received enquiries about firewood as fuel from various countries.

Apart from these, our lime makers are also using charcoal from wood, coconut shell and cashewnut husk, coal cinders, coke and lignite. The results of these alternative fuels are very satisfactory.

Lime pozzolana mixture (KVIC-LYMPO)

After starting lime production in 1961, KVIC began work on developing a cheaper alternative to cement. One such product is a traditional lime-sand-surkhi mortar (lime, sand and brickdust) to suit the rural requirements for building construction in structural concretes. Previously mortar was produced in India on a very limited scale using the bullock-driven ghani method. This was the best binder before the advent of Portland cement. Most Indian monuments show evidence of the use of lime mortar (such as dams, bridges, monuments, buildings, forts, railway bridges, tanks and canals).

The introduction of LYMPO production was the result of a techno-economic survey by KVIC and the technology was developed by the organization because there was an acute shortage of cement. LYMPO production was introduced all over the country as a cheaper binding material and substitute for cement. LYMPO is a mixture of lime and burnt clay and is known as KVIC's pozzolana.

LYMPO is a masonry binder with high plasticity and workability, low shrinkage, high water retentivity and low porosity.

The main raw materials used are:

- calcareous materials such as lime/limestone/lime shell, and
- a pozzolanic material to provide the silica and alumina and so improve upon the hydraulic and cementation indices.

Limestone/lime shell

Lime shells of around 99 per cent $CaCO_3$ and limestones at least 95 per cent are calcined under KVIC specified conditions to produce quicklime. Fat, lean or hydraulic varieties of lime are produced depending on the type of raw material.

Burnt clay pozzolana

The common clays which conform to the following chemical analysis and are free from lumps and coarse gravel can be moulded into bricks for calcining to the optimum temperature in a brick bhatta.

The optimum temperature of burning is that at which the crystal struc-

Constituents	Per cent
SiO_2	>45
$SiO_2 + Al_2O_3 + Fe_2O_3$	>70
CaO	<8
MgO	<3
Soda and potash	<3
LOI	<5

ture of the clay mineral just collapses and the oxides of silicon, aluminium and iron are in the fine active state and thereby maximum reactivity is achieved.

The occurrence of weak bricks in a bhatta is seen in the top layers or extreme corners (where the temperature falls below 800°C), and as a result they cannot be properly fired due to this remoteness. Thus little or no strength develops and they are not used in construction. On an average, a brick bhatta of two lakhs bricks provides around 40–50 tonnes of such waste material which serves the purpose of a burnt clay pozzolana. The lime reactivity obtained from this type of pozzolana ranges from 450 to 550psi (30–40kg/cm).

Further, in a continuous operation such pozzolanas can also be manufactured in an improved vertical kiln with a grating and induced draught arrangement giving better control for burning the clay brick pieces at 800–850°C in a mixed-feed process, as designed by KVIC.

In view of the absence of clay components in fat lime having 65 per cent CaO, its hydraulic and cementation values are poor. On average, it can absorb nearly twice its weight of the burnt clay pozzolana to improve its binding properties through the addition of SiO_2 Al_2O_3 and Fe_2O_3 components.

Gypsum

To accelerate the setting time so that the speed of construction is faster 4–6 per cent mineral gypsum, in powdery form, having $CaSO_4$ – 70 per cent with low P_2O_5, is used. This not only enhances the initial setting time but has been shown to improve the early strength.

Production method

Processing of raw materials

The limestone or shells are burnt at their dissociation temperature by feeding the required quantity of solid fuel and air draught in the KVIC kiln which is not only economical to use but also easy and continuous in operation, and produces well burnt and consistent quality lime. The burning percentage is never below 96 per cent. The quicklime thus obtained is hydrated by either a platform or a chamber process (as evolved by the Commission) by adding a calculated quantity of water and recirculating the steam generated during the exothermic reaction with the object of meeting the chemical requirement of quicklime (CaO) for water. The burnt clay pozzolana is disintegrated and pulverized to a fineness of 1.18mm and the hydrated lime is screened through 15 mesh (1.18mm I.S.) separately. Then they are batched by the specific ratio by weight as per the grade of lime. The requisite quantity by weight of mineral gypsum powder is then added.

Mixing operation

The batched material is manually mixed on a platform before putting into the machine. The small lumps if formed should be broken by hard pressing. It can then be passed on to the intergrinding stage.

Intergrinding

To obtain a homogeneous mass and size reduction, the materials are passed through two roller mixers. The rollers should have a minimum weight of 100kg each and run at 120rpm for 10–12 minutes.

Pulverization and packaging

The product thus obtained is carried to the pulverizer connected to the mixer through a channel so that the end product is further reduced to 125–106 micron (120–150 mesh).

A polythene-lined bag is attached at the outlet of the pulverizer so that the product can fall directly into the bag, on a platform balance. The filled bag, weighing 50kg, is stitched up before despatch.

Advantages of KVIC lime kiln technology

1. Increased height with refractory lining.
2. Works on natural or controlled draught.
3. Works continuously and can be adapted for the quality of lime required.
4. Produces a uniform quality of lime.
5. Better thermal efficiency.
6. Saves fuel, thus reducing cost of production, resulting in higher margins and better wages to artisans.

Conclusion

KVIC offers technical guidance, technology design, feasibility surveys, erection and commissioning of plants on a turnkey basis. Lime technology has been transferred to other countries where the results are very satisfactory.

KVIC has set up 5615 institutions and 9683 co-operative societies, and 172 412 individual entrepreneurs have availed themselves of either technical or financial assistance during the period from 1963 to 1990, generating employment opportunities for more than 6 lakh persons, many of these in the field of lime and alternative cements.

21. Lime production in Algeria and future prospects

J. MOUSSA*

The production of binders is of great importance in developing countries. They are the essential components of all construction and the cost of binders represents a very large expense for low-income communities, even for their governments which are required to import great quantities of these materials. Because of the new price of cement in Algeria, which has seen a rise of 400 per cent, it became necessary to promote local production of economical binders (alternative cements).

In developing countries, the deployment of the cement works industry has been carried out intensively to the neglect of traditional lime and gypsum factories which exist in many regions. In Algeria and Tunisia, for example, all the traditional lime kilns are disappearing, even though lime was the favoured binder for almost four thousand years, for preparing mortars and even for the waterproofing of roofs and pipes.

The great wall of China, the Coliseum of Rome and the city of Ghardaia in Algeria are evidence of the technical mastery of our ancestors in the use of lime.

The state of lime production and lime-based building materials in Algeria

Within the framework of the development of local materials, an interdepartmental group[1] has been set up in order to co-ordinate the main line of construction techniques development. Lime, calcium silicate bricks and autoclaved aerated concrete are listed among the local materials proposed by this group. The predictions for these three materials are presented in Table 1.

---

*CNERIB, Algeria.

Table 1. Production predictions for 1990

Item	Volume in 1984	Vol. predicted for 1990	Number of factories for 1990	Cost (millions of Algerian dinars)*
Lime factory	178 000 tonnes	1 800 000 tonnes	8	1 900
Autoclaved aerated concrete factory	125 000 m³	1 000 000 m³	8	1 800
Calcium silicate bricks factory	0	1 000 000 m³	10	1 200

* 40 Algerian dinars = £1 sterling.

In fact, in 1991 the achievements are far from the predictions; this is shown in Tables 2, 3 and 4. By comparing Tables 1 and 2, one can see that the increase in lime production predicted by the group has not materialized.

Table 2. Lime factories existing in 1991

Factory site	Capacity (tonnes per year)	Type of kiln
Berriane (Ghardaia)	17 000	rotary
Bouinan (Blida)	20 000	shaft
Chettaba (Constantine)	20 000	rotary
Metlili (Ghardaia)	21 000	shaft
Oum Djerane (Saida)	100 000	rotary

Total capacity: 178 000 t/y

Table 3. Autoclaved aerated concrete factories in 1991

Factory site	Volume (m³)	Observations
Meftah (Alger)	125 000	working
Oum Tboul (Tebessa)	125 000	production stopped

Total capacity: 250 000 m³

Only two of the eight predicted autoclaved aerated concrete factories have been constructed.

Table 4. Calcium silicate brick factories

Site	Volume (tonnes per year)	Observations
Bou Saada (M'Sila)	140 000	in production since Nov. 1991
Naama (Ain Defla)	70 000	in production by Feb. 1992
Relizane	210 000	″ ″ by May 1992

Total capacity: 420 000 t/y

Table 5. Comparison between lime and cement prices in Algeria

Type of boiler	Factory site	Price per tonne (Alg dinars)	Year
Portland cement	Meftah (Alger)	300	1987
	"	600	1990
	"	1 600	1991
Quicklime	Bouinan (Blida)	480	1991
Hydrated lime	"	560	1991
Hydraulic lime	Chettaba (Const.)	500	1991

Only three out of ten calcium silicate brick factories have been built. The reasons for this delay are first of all financial (slump in the oil price) and, secondly, the subsidizing of cement by the Algerian government to the detriment of gypsum and lime. The government subsidy ceased only in 1991. A comparison between lime and cement prices is presented in Table 5.

Means of lime production in Algeria

There are two principal types of lime kiln in Algeria, shaft and rotary kilns. A third type is the traditional kiln. The industrial kilns (shaft, rotary) tend to be great concentrated factories with high nominal capacities (100 000 tonnes/year, minimal break-even point) and heavy investments. These kilns are not appropriate for Algeria for the above reasons, i.e. high nominal capacity and heavy investment. We should build intermediate capacity kilns, appropriate for local production requirements. This type of factory (from 2500 to 25 000 tonnes/year) allows a better geographical distribution of production, and reduces transportation problems and stock shortages in case of breakdown. Using an easy controllable technology, it makes feasible the use of existing local infrastructures.

It is advisable to build this type of factory in the following cities:[2] Ain Sefra (10 000 tonnes/year), Bou Saada (20 000 tonnes/year), Relizane (25 000 tonnes/year), Tebessa (15 000 tonnes/year), because of the calcium silicate bricks factories built in these regions.

Suggested process to promote this material[3,4,5,6,7]

Diagnosis of existing factories
Before building other lime factories, it would be better to report on the existing ones. This report will help the understanding of the problems faced in these factories and to choose the best sites and capacities of future factories.

2. Technical basis of the different uses of lime
In order to establish the technical basis of certain lime uses, CNERIB is establishing for each use: the state of knowledge, experimental

programmes and guides and standards. The envisaged uses of lime are limited to the construction sector for joint mortaring, coating, rehabilitation works, painting and plasticizing.

3. Pilot application of the uses of lime
A prototype house made from calcium silicate bricks will be built in CNERIB and joint mortar and coating made from lime will be tested on it.

4. Information and training
In order to promote products, and for popularizing the use of lime or construction elements made from lime (autoclaved aerated concrete, calcium silicate bricks), the following actions are necessary:

- Information, including conferences and seminars; audio-visual documents and films; TV programmes, and papers.
- Cycles of long-term training and introducing these materials into courses for training architects.

Conclusions

The construction programme is increasing dramatically in Algeria, therefore it is necessary to widen the possibilities or means at a lower cost, i.e. exploiting natural resources of calcium carbonate. The development of building materials needs to go hand in hand with:

- study of deposits and markets before the choice of sites
- mechanization of the traditional means of production
- development of permanent maintenance in the mechanized factories
- training of factory workers.

International co-operation will be needed to help in the development of means of production appropriate for Algeria.

With the demand of the calcium silicate brick factories in 1992, and applying the true price of cement, a shortage of lime in 1992 is predicted.

22. Technology option for manufacture of calcined clay pozzolana (surkhi)

J SEN GUPTA*

Pozzolanas are siliceous material which, while having no cementitious value within themselves, will, in finely divided form and in the presence of moisture, chemically react with calcium hydroxide at ordinary temperatures to form compounds possessing cementitious property. Surkhi (burnt clay) is one of the common pozzolanic materials used in India from olden times. In most of the ancient constructions which are still standing today, surkhi was used in conjunction with lime. At present, surkhi is produced in the country mainly by grinding brick bats. It is of variable quality and generally of poor pozzolanic activity. Such surkhi, when used with lime for mortars, has very low strength. With the availability of cement which gives early strength, the use of lime-surkhi for mortars has been considerably reduced in building construction, particularly in urban areas.

Investigations carried out by the Central Road Research Institute (CRRI), New Delhi, and other research institutions in the country, have shown that certain type of clays, when calcined at optimum temperature, give rise to good pozzolanic reactivity. The clay pozzolanas (reactive surkhi) thus prepared when used as lime-pozzolana-sand mortars have the same strength as cement-sand mortars. The National Buildings Organization (NBO), in collaboration with Central Road Research Institute, conducted a survey of clay available in various parts of the country, which could be used for manufacture of standard quality clay pozzolana.

About 30 per cent of cement consumed in building construction is for mortars and plasters. The clay pozzolana produced under controlled conditions in conjunction with standard quality lime could form a good substitute for high energy and capital intensive cement for mortars and plasters. The clay pozzolana replaces ordinary Portland cement (OPC) to the extent of 20 per cent in cement concrete and mortars without affecting the strength and it also constitutes up to 25 per cent of pozzolana cement.

The advantages of clay pozzolana when used with lime or cement in the preparation of mortar and concrete are:

- improved workability
- lower heat of hydration and thermal shrinkage
- increased water tightness

---

*Building Materials & Technology Promotion Council, Ministry of Urban Development, New Delhi.

- improved resistance to attack by sulphates in soil and water
- reduced alkali-aggregate reaction.

Present status

Clamp kilns

In Madhya Pradesh, Maharashtra and Rajasthan, locally available plastic clays are moulded into balls, dried and fired in a clamp kiln using coal cinder or any other fuel such as cowdung. The balls are crushed and finally ground in grinding mills. The pozzolana produced in the clamp kiln is of variable quality and is used mainly in the construction of buildings in rural areas.

Down-draught rotary kiln

The process developed by CRRI for production of clay pozzolana uses down-draught or rotary kilns, depending upon the scale of manufacture. The clays, either in briquette or pellet form, are calcined at a predetermined temperature. After proper calcination, the material is finely ground in a ball mill or a tube mill, so that 90 per cent of it passes through an IS–15 sieve. A demonstration-pilot plant was set up by CRRI and over 1500 tonnes of calcined clay pozzolana and lime pozzolana was produced between 1963 and 1967, and supplied to various construction agencies.

Shaft kilns

Gujarat Engineering and Research Institute have developed a vertical shaft kiln for production of clay pozzolana. In this process, the feed consisted of a mixture of clay lumps 50–100mm in size and coal slag. The conditions of calcination were 700°C for three hours. This was maintained with the aid of a thermocouple and controlled via an airblower and feed input.

NBO has developed a process for the production of clay pozzolana by a fluidized bed technique which has higher thermal efficiency. The laboratory investigation, pilot plant studies carried out at Sri Ram Institute for Industrial Research (SRI), Delhi and the large-scale production unit put up by NBO are dealt with below.

Fluidized bed technique

Fluidization is a physical phenomenon whereby a bed of granular material agitated by a current of gas passing through it exhibits some of the properties of a fluid. This method of gas/solid contacting has a number of characteristic advantages, such as enhanced heat and mass transfer rates, uniformity of temperature, ease of handling solids and adaptability to continuous operation. Due to these advantages, the fluidized bed technique has been widely adopted in the petrochemical industry for carrying out gas/solid catalytic reactions. Other applications are in the granulation of fertilizers, manufacture of organic chemicals, mixing of fine powder coating of plastics on metal, etc. The technique has been used for the calcination of limestone and commercial units for the production of lime are in operation.

The experimental investigations that have so far been accomplished on the degree of calcination of clays for manufacture of surkhi pozzolana were confined to heating the material in muffle furnaces, down-draught kilns, country kilns and rotary kilns. The mode of heating, particle size of clay and the retention time of the material inside the calciner, are some of the important factors which govern the economics and the quality of the product. In the fluidized bed techniques for heat activation of clays, the clay is handled in the form of fine particles and there is good intimate contact between the fluid (hot gas) and the solids (clay particles) in the system. The retention time of the particles is low. All these factors assist in getting a product of uniform quality and better heat transfer, thereby reducing the cost of production. With this in view, a research scheme on the design and development of fluidized bed calciner for production of clay pozzolana was undertaken at SRI.

Initially, laboratory studies were undertaken at SRI to determine flow characteristics of the air-solid system with emphasis on solid rate, gas rate and solid hold up in the column. Experiments were carried out in a 3 inch (75mm) square section plastic model with a height of 4ft 3in (1.3m). Fins were arranged in the column in order to achieve uniform contact of gas and solid. With the help of sand particles at a predetermined rate, the choking velocities were recorded for all experimental variations of particle size, solid rate and air velocity. A rotameter was used to measure the flow of air. Solid hold up in the column was calculated, using the following equation:

$$W = P_p (I - E_s) (V_e - V_{ch})^4,$$

where

$W = \dfrac{\text{lbs of solid}}{(\text{second}) (\text{square ft})}$ of the column

P_p = particle density $\dfrac{\text{lbs}}{\text{cu. ft.}}$

E_s = voidage, dimensionless

V = velocity ft/sec at voidage E_s which is equal to free fall velocity or terminal velocity of the particle (calculated by applying modified stockes law equation)

V_{ch} = choking velocity, ft/sec.

Table 1. Flow characteristics of air-solid system in the plastic model

W (lb/(sec)(sq ft))	Vch (ft/sec)	Solid hold-up (I-Es)
2.3	16.0	0.028
8.0	13.3	0.023
11.4	12.0	0.020
18.5	10.6	0.023
38.5	66.6	0.026

Table 2. Results of studies on the flow pattern of air-solid system

Experiment No.	Material	Volume of solids	Solid rate (lb/min ft)	Air velocity	Hold up solids
1	Glass beads (–60+80) mesh	30	99.0	1.15	0.58
2	"	40	99.0	1.73	0.775
3	"	62	99.0	2.19	1.20
4	"	70	99.0	2.85	1.36
5	"	75	99.0	3.10	1.45
6	"	96	99.0	4.75	1.86
7	Sand	30	77.6	1.85	1.55
8	–60+80 mesh	40	77.6	2.4	1.73
9	"	60	77.6	3.1	1.16
10	"	110	212.0	4.75	2.13

The results are given in Table 1. From the results, it appeared that the solid hold-up in the column is around 2 to 3 per cent. Further studies on the flow pattern of the air solid system were conducted in the experimental unit by providing two slits in the fluid bed section at a distance of 3ft apart, to confirm the validity of this equation. Solid hold up in the column was calculated from the following expression.

$$I - E_s = \frac{\text{volume of solids retained in the test section}}{\text{volume of the test section}}$$

The result of this study is given in Table 2. The results of the flow characteristic study of air-solid system in counter-current fluid bed unit indicate that a maximum solid hold up of 2 to 3 per cent can be achieved without choking or slugging of the solid particles in the column.

Development of a fluidized bed calciner

The data available from these studies were utilized for designing and fabricating a pilot plant unit having a capacity of 2 tonnes of clay pozzolana per day. This unit was constituted of the following:

- dilute-phase counter current fluidized bed calciner

- pneumatic transport for solids
- oil burner and blower system
- thermocouples and the flue gas analyser.

The calciner is divided into the following zones:

(a) Fluid bed column
(b) Burning zone
(c) Product outlet chute
(d) Fine particle recovery section.

In order to improve the heat transfer efficiency and to achieve uniform gas heat contact, fins were provided in the fluid bed section of the calciner.

The counter current system has got an advantage over the conventional fluid bed system since it could be operated at a higher fluid velocity than the conventional fluid bed system. In the dilute phase, the concentration of solid in the gas-solid mixture never goes beyond 4–5 per cent, which ensures a smooth gas-solid contact and eliminates the formation of bubbles and slugs.

The unit was utilized for a continuous long duration trial production of clay pozzolana to determine the rate of output, consumption of fuel, uniformity of the product and to have an idea about the cost of production.

Raw materials

Selection of the right type of clay and optimum temperature of burning are important factors for the production of pozzolana of high reactivity. The chemical composition of the clays recommended for production of pozzolana are given in Table 3.

The raw material used for pilot plant studies was a ferruginous clay, rejected materials from kaolin mines at Mahipalpur 20km from Delhi. The optimum temperature of calcination is determined by calcining it in the laboratory at different temperatures.

Table 3. Characteristics of clay suitable for production of calcined clay pozzolana as per IS: 1344–1981

Constituents	Contents
Silica+Alumina+Iron oxide ($SiO_2+Al_2O_3+Fe_2O_3$)	Not less than 70 per cent
Silica (SiO_2)	Not less than 40 per cent
Calcium oxide (CaO)	Not more than 10 per cent
Magnesium oxide (MgO)	Not more than 3 per cent
Sulphuric anhydride (SO_2)	Not more than 3 per cent
Soda and potash (Na_2O+K_2O)	Not more than 3 per cent
Water-soluble alkali	Not more than 0.1 per cent
Water-soluble material	Not more than 1 per cent
Loss on ignition	Not more than 10 per cent

The clay pozzolana produced from the pilot plant at the optimum temperature had a lime reactivity value of 70 to 90kg per cm^2.

The pilot plant investigations have established that: (a) the fluidized bed technique could be effectively used for the heat activation of clay for production of uniform quality of pozzolana; (b) higher thermal efficiency of the process enables considerable saving in the consumption of fuel as compared to other processes; (c) as there are no moving parts, the calciner unit requires little maintenance as compared to rotary kilns; and (d) the final grinding cost of the finished products is less since the materials which come out of the calciner are finer than 10 mesh.

Scaled-up unit

Based on the data obtained from the pilot plant studies, a demonstration-production plant with a capacity of 20 tonnes per day of clay pozzolana was set up by NBO at Sultanpur, Delhi at a total investment of Rs6 lakhs (US$24 000). The main objective of this unit was to make available standard quality clay pozzolana to construction agencies in Delhi and to impart training to the new entrepreneurs desiring to set up such plants in the country and also for test evaluation of the raw materials. The total height of the calciner is 35ft (over 10m) with a cross-section of the fluidized bed column 30cm × 30cm. The ceramic fins have been replaced by metallic stainless steel fins. An additional layer of insulation bricks has been used in the burning zone.

The process

The dried clay lumps are disintegrated in a disintegrator down to the size of 10 mesh and stored in the clay pit below the ground level. The clay is carried from the pit by means of the skip hoist to the hopper placed at the top of the calciner, from where the clay is discharged through a screw feeder into the calciner. As the material falls due to gravity and comes in contact with the up-draught of hot flue gases generated by the oil burner, the intimate contact between the hot gas and clay materials is ensured by providing metallic fins inside the calciner which obstruct the passage of the clay material and retain it for a few seconds. The material is then passed through the burning zone and comes out of the exit chute where it is allowed to cool for 24 hours. The cooled material is pulverized in a pulverizer to pass through BS100 mesh sieve. The finished material is then bagged, stitched and stored.

Energy requirements

Ordinary surkhi is prepared from the rejected brick bats by disintegrating or pulverizing them. Hence, no energy should be considered to have been used in the process because clays are fired for obtaining bricks. However, energy input required for pulverizing these brick bats has been estimated as 50Kcal/kg.

The energy requirement for calcination of clay based on the fluidized bed process where fuel oil is used has been worked out to be 228Kcal/kg per calcination and about 56 Kcal/kg in the form of electrical energy for grinding.

The total energy requirement for the production of lime pozzolana mixture (1 hydrated lime:2 calcined clay pozzolana) conforming to Indian Standard Specifications has been estimated to be between 343 and 423Kcal/kg with the thermal efficiency of the lime kiln ranging from 60 to 40 per cent.

Based on the above data, the energy requirements for mortars of various types have been estimated and are given in Table 4. It may be seen that mortars containing cement are high energy consuming. There is a considerable saving in energy in the use of lime-pozzolana mortars as compared to cement-sand mortars.

Table 4. Energy requirements for mortars/plasters using lime/lime pozzolana vis-a-vis mortars using cement

Binder	Binder: Sand	Energy consumption for binder in terms of 1000Kcal/m for wet mortar	Average compressive strength (28 days) kg/cm^2
Hydrated lime fly ash	1:0	150–230	16.0
Hydrated lime fly ash based activated LP mix	1:2	230–350	21.0
Hydrated lime burnt clay pozzolana	1:3	110–140	30.0
Activated lime burnt clay pozzolana based	1:3	110–140	30.0
Rice husk ash masonry binder	1:3	40–60	45.0
Portland cement	1:6	215–280 (D) 335–390 (W)	35.0 35.0

(W) Wet process   (D) Dry process

Conclusion

A clay pozzolana of uniform quality conforming to IS Specifications could be produced by fluidized bed technique. The optimum economic capacity recommended for the process is 20 tonnes per day. There is good scope for the setting up of such plants in various parts of the country where deposits of suitable clay and limestone are available. The main advantage of the fluidized bed process is in the simplicity of its operation resulting in low maintenance and repair cost. The energy consumption compared to other processes is also less and provides uniform quality of product at a lower cost.

SECTION III

Uses of lime

Small-scale lime production in developing countries has received a marked increase in attention since 1974. There is now a greater awareness of the potential for substantial growth of the lime industry in developing countries, coupled with an increased understanding of the importance of lime for construction.

The successful development of small-scale lime production with which ITDG has been associated in countries such as Malawi and Zimbabwe over recent years has been remarkable. There is an economic rule, however, which must logically be followed. This is that supply follows demand. Provided the manufacturing, agricultural, and construction industries are sufficiently advanced, the demand for lime will be immense.

As industry expands in these countries and the requirement for lime increases the type, quality, and frequency of demand will fluctuate. One reason for the success of recent small-scale lime production in Uliwa and Balaka, Malawi is the achievement of a product closely matched to the precise demands of the local sugar industry.

The demand for lime in developing countries at present tends to be limited to specific industries such as agriculture, the production of fertilizer, soda ash, chemicals, and sugar refining, which are usually served by their own captive lime production units. Lime is, however, also an entirely appropriate binder for localized and low-cost construction throughout most of the world.

Traditionally, before the advent of Portland cement, lime of a relatively consistent standard was in constant demand by the building industry.

The following chapters concentrate, therefore, on the development of lime for the construction sector. It is clear that this is a subject fundamental to progressing both low-cost and self-sufficient means of providing shelter, and improving living conditions in the many underprivileged regions of the world.

There has been a parallel resurgence in the research, development, and use of lime for the conservation of historic buildings, principally in Europe but also for the care and repair of historic buildings in developing countries, including world heritage sites recognized by ICOMOS.

In the past, the widespread use of lime for building led to the need to improve the performance of lime mixes to withstand a variety of environmental conditions, and to suit the requirements of different building elements. Performance was improved by incorporating additives which may be broadly summarized as:

- **Aggregates**: sands, stone dust, grit, and inert soils, suitably graded.
- **Pozzolanas**: brick dust, volcanic and fuel ashes.
- **Organic additives**: keratin, sugar, fats, and oils.
- **Setting agents**: these assist workability and/or set, such as gypsum, size and salt.
- **Fibres**: mostly to assist bond and tensile strength which include a range of animal hairs, cowdung and vegetable fibres.

In addition to many alternative additives, the methods of production, storage, preparation, mixing, and application all have an effect on final performance. Generally the methods and techniques used, although having broad principles in common, can best be developed with local experience and demonstrated by craftsmen practised in the use of these materials.

Variations in the preparation of lime mixes and use are immense in England alone. There are many examples and records of their successful use in Europe, although much research into the method of application remains to be carried out. Detailed records of lime mixes for early buildings in developing countries are scarce, and extensive archaeological research is still required. The permutations of various materials and mixes used in association with lime are almost infinite.

Demonstrator specifications

Lime mix specifications successfully used in the past and currently used for carrying out matching repairs to historic buildings are detailed in Appendix 1 by J. Orton.

Organic additives

Current research into some organic additives and their effect is described by C. Santiago, and the current three-year Smeaton Research Programme is detailed by I. McCaig.

Aggregates

Research and development of waste products, particularly fuel ash, as aggregates is described by B. Baradan, T.A. Schilderman, and N.G. Dave.

Soil stabilization

The use of lime for soil stabilized building components and construction is covered in detail by D.J.T. Webb and H. Houben.

Building conservation

The current use of lime for the care and repair of historic buildings is described in a series of examples by J. Bucknall for work in the UK, and by F.I. Kara in Zanzibar. J. Fidler sets out initial co-ordination arrangements for technical research and development on building limes in connection with the repair of historic buildings in Europe.

Socio-economics

The introduction and expansion of localized lime production and use has marked socio-economic implications which are discussed by N.G. Dave and T.A. Schilderman.

Standards

Resurgence of the production and use of building limes will necessitate a generally accepted means of testing and establishing its quality. This is clearly essential if confidence in the product is to be maintained for the benefit of both consumers and producers alike. Complex chemical analysis is not appropriate for small-scale local production and use. This will often be in areas where access and transport are limited. M. Wingate sets out initial proposals in respect of appropriate standards for building limes in this context, a timely reminder of the importance of arriving at appropriate internationally accepted standards for building limes in the near future.

S.D. Holmes

23. Organic additives in Brazilian lime mortars

CYBÈLE CELESTINO SANTIAGO* and MÁRIO MENDONCA DE OLIVEIRA*

We agree with Ashurst[1] when he states that lime is one of the most neglected building materials in the world. In Brazil, as in other countries, the popularity of Portland cement, even if just a fashion, caused the abandonment of lime in construction. If lime were not fundamental to metallurgical industries, agriculture, soil stabilization, potash, mineral paint production and other activities, its industrial production would have practically disappeared, even though we are among the largest producers of this material in the world. From our point of view, although lime lost its place as the most common binder in construction it should recover at least part of its status in the building industry. Lime mortars are usually better than cement mortars, not only for new construction, but particularly when used in the conservation and repair of historic buildings.

Compatible materials are required for the repair and conservation of old structures, plasters, and renders to achieve continuity and avoid colour and texture changes, or undesirable secondary effects of joining dissimilar materials. There is a possibility of soluble salt migration in cements, mainly if Portland cement is used; consequently cement is generally not recommended for repair as the migration of salts within the structure is very dangerous. In Brazil, and in other countries of Latin America, the restoration of earthen structures is common. In these structures such as adobe, lath-and-plaster, or *pisé de terre*, lime is normally specified as the additive in connection with all repairs.

There are very interesting experiences regarding the use of lime as soil stabilizer in Bahia, Brazil. For example, lime has been combined with bamboo piles in the construction of railway foundations, a very common practice, especially if the soil is *massapê*. These materials are abundant in our area. Soils are very rich in montmorilonitic clay and present many stabilization problems. We are aware that some Latin American countries have a traditional method of reinforcing the foundations of existing structures by filling peripheral holes with lime. There is a need for further research on this subject. There is much to be learned from archival research and recovery of information now lost but once common knowledge.[2]

---

*Univeriadade Federal da Bahia, Brazil.

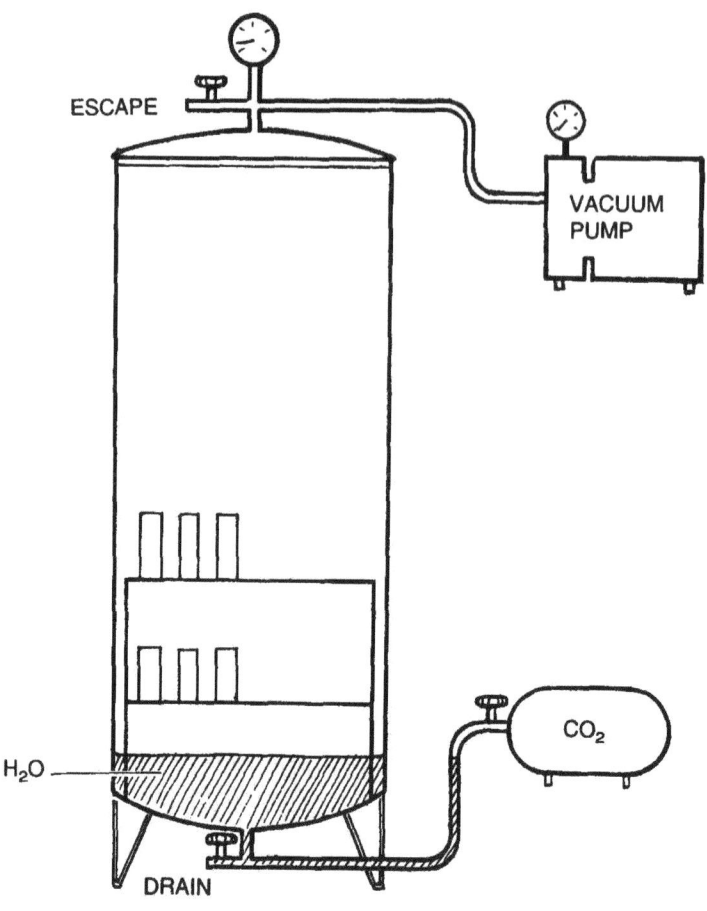

Figure 1: *Accelerated carbonation chamber*

In Brazil, even if the tradition of manual lime production is no longer continued, a return to this method would not be difficult due to the many possibilities of using materials that are already proven. Nevertheless, there is still room for research due to the wide range of uses for which lime is suitable. There are continuing efforts by many organizations to find alternative technologies for building low-cost units in developing countries. Although the production of lime is industrialized at present it does not preclude manual processes. The same is not true for Portland cement production. The possibility of small-scale lime production can benefit those low-income groups who live in limestone rich areas, or near lime formations of biological origin, that were commonly used in the past.

There are institutions in Brazil that have carried out intensive research regarding the use of lime in low-cost buildings. Among these we would like to point out work developed by the *Associação Brasileira de Productores de Cal* (ABPC, Brazilian Association of Lime Producers), *Instituto de Pesquisas Tecnológicas* (IPT, Institute of Technological Research) and *Centro de Pesquisas e Desenvolvimento* (CEPED, Research and Development Centre). Research is still needed to explore the cultural and historical traditions of the material. We also need to develop laboratory techniques that facilitate this work as well as the research that must be pursued. One area to progress is to recover and understand lost traditions and to determine where unsatisfactory variations or distortions have occurred when knowledge was transferred. It is also necessary to discover the reasons why the use of specific technologies for lime production has been discontinued.

Our first step in the study of lime mortars was to develop a faster way of achieving results. We know the slow process of carbonation of lime mortars; there are even proverbs that state: 'lime at a hundred years old is still a baby and when a hundred years are past and gane, then gude mortar turns into stane' [sic].[3] These statements explain the problem quite well.

All researchers know that the carbonation process can be accelerated by putting the mortar in contact with CO_2. We developed an efficient method of carbonating the interior of samples by using a vacuum followed by the creation of a CO_2-rich atmosphere (Figure 1). The first results were very good.

The second step was to use different techniques to analyse old mortar. This is a controversial subject. Our objective was to find the proportions and components of the mixtures.

In the third step we concentrated on studying organic additives in lime mortars. The preliminary result of this research constitutes the principal topic of this paper.

Organic additives

According to oral tradition it was common practice in the past to include whale oil in Brazilian lime mortars. The main reason for this would be the exceptional properties of the oil to increase mortar strength. This would also explain the survival of many historic buildings to the present day. Contrary to this, however, no evidence has been found in national literature to support the theory of using whale oil for this specific purpose. There is little known reference to the product in any list of building materials in Brazil or any other parts of the world[4], and its use is rarely mentioned. References that are available are not clear, making it impossible to reproduce the proportions. Furthermore, it has not been possible to establish the validity of the information we have been given.

Considering the oral tradition alone, the number of buildings assumed to be made with whale oil was so great that there would not have been enough whales to produce the quantity required. It is said that historic buildings existing inland were built with a whale oil additive in the mortar. We question this assumption due to difficult transport over considerable distances increasing the cost of construction. How could our oil supplies be enough for this as well as for the production of many common goods such as public illumination, soap, caulking, candles, and the leather industry?

To verify whether whale oil additive was really used in our mortars, we prepared some samples.[5] We submitted these samples to different tests – capillary absorption, strength, weathering by salt crystallization using Na_2SO_4 – in order to verify the behaviour of the mixtures. The absence of clearly defined mortar formulae that included whale oil led us to use arbitrary mix proportions. Thus, we chose 1:3 (lime, sand) and 1:2:1 (lime, clay and sand) as parameters to avoid much variation. The quantities of oil added were 5 and 2 per cent, according to the weight of lime. We experimented with 10 per cent oil, but the mixture did not bond well, probably because the proportion of oil was too high.

We began the experiment with capillary analysis. Our aim was to test whether the oil was able to give any hydro-repellance to the formulae.[6] In general, the samples were 240 days old when submitted to that kind of

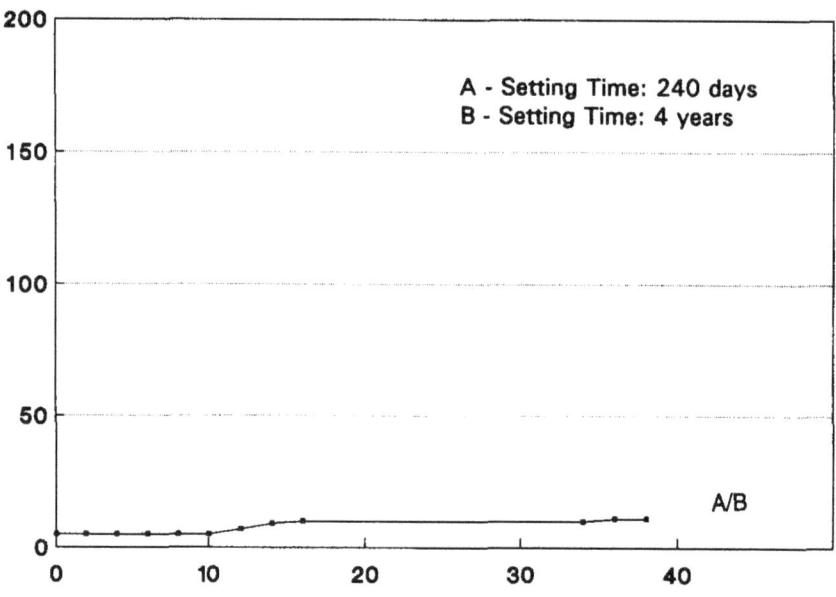

Figure 2: *Capillarity absorption of mortar with whale oil*

analysis. We also tested the behaviour of a four-year-old group of samples. The results may be observed in Table 1 and in Figure 3. Four-year-old mortars with high oil content (5 per cent) were more impermeable than the ones with less oil (2 per cent) and than those with no oil. In the case of four-year-old mortars with 5 per cent oil content the degree of water absorption was very low. The graph curves we obtained almost coincided with the horizontal axis. This result also proves that even after four years, the oil had not lost its ability to improve impermeability.

Besides corroborating old assumptions, the results ratified the information that an elderly mason gave us: he told us he used whale oil as an additive to lime mortar for the construction of a quay to improve impermeability. He also pointed out that he only used whale oil because it was abundant on Itaparica Island, where he was working, not because the oil was different, or unique.

In his opinion, any oil would be suitable as a bonding agent. On the other hand, the results of the strength tests proved that mortars with an oil additive are less resistant to compression than regular mortars, probably because the oil encapsulates the $Ca(OH)_2$ particles and prevents the carbonation process taking place. We have, therefore, arrived at the conclusion that the traditional belief is not fully justified. Possibly if we had used oil (coarse) stuff, as Ellis indicated,[7] another factor would have influenced the mortar's performance. Unfortunately, whales are now rare in Brazil and so are *armações* (places to transform whale's lard into oil). For this reason, it was impossible for us to acquire the material and to test it.

The addition of sugar to lime-mortars is still common practice in Brazil when the mortar is to be submitted to high temperatures. We know that sugar has been used in different forms in various areas of construction when a weather-resistant mixture was necessary.[8] Consequently, we decided to carry out a weathering analysis of mortars prepared, by simulating the effects of weathering and salt crystallization. Our samples were submitted to consecutive dry-humid cycles in a saturated solution of Na_2SO_4 in water. This is normal practice for weathering analysis. At the end of the fourth cycle, the mortar, soil and lime samples without oil already showed signs of decay.

At the end of the sixth cycle, all of them were completely destroyed, except the ones that contained sugar. Besides resisting the washes very well, they also resisted damage from efflorescent salts. This confirmed that the inclusion of sugar increased weathering resistance of the mortar as well as its resistance to salt efflorescence, and frosting. We had better results in strength tests on mortars that contained sugar than on additive-free mortars (Table 1). The samples that contained oil gave the worst results.

Capillary absorption analysis demonstrated that samples with the highest amount of sugar[9] resisted water penetration better than the ones with less of the same additive, the latter giving better results than ordinary mortars,

Table 1. Compression strength

Mortar	kgf/cm²
S+L (additive free)	11.0
S+L+O (2%)	4.3
S+L+O (5%)	3.5
S+L+Su (11.9% H_2O)	26.7
S+L+Su (25% H_2O)	25.8
S+L+So (additive free)	20.9
S+L+So+O (2%)	18.5
S+L+So+O (5%)	6.3
S+L+So+Su (11.9% H_2O)	22.1
S+L+So+Su (25% H_2O)	35.3
S+L+So+B (25% lime)	14.3
S+L+So+B (50% lime)	20.9

but not as much as the samples that contained oil. We submitted the mortars that contained sugar to heating. Through this analysis we proved that up to 600°C the samples did not show any visible change. At temperatures higher than 900°C, the mortars without an additive broke apart. The same happened to ordinary mortars that contained sugar. Ordinary

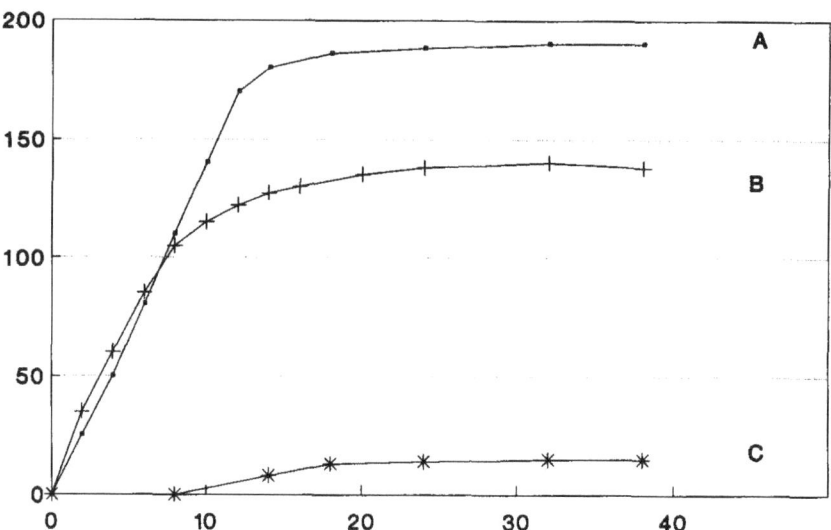

A - Sand, Lime Mortar without additives
B - Sand, Lime + Sugar
C - Sand, Lime + Whale Oil (5%)

Figure 3: *Capillarity absorption – comparative curves*

'bastard' mortars (mortars with clay) displayed some cracks, but sugar-containing ones presented no visible problems (Figure 3).

The mortars that contained sugar weighed less when submitted to high temperatures, probably due to the decomposition of the sugar. Our goal now is to carry out mercury porosimetry analysis of all tested samples in order to verify any modifications in porosity due to the different additives we used. Unfortunately, there are many problems regarding scientific research in Brazil. We have no means of buying expensive equipment, nor repairing the old equipment we have. For this reason we decided to try another experiment, cheaper than the IR spectroscopy, or the GC and FTIR analysis.

The sugar identification in our old lime mortars was done through Benedict's test as described by Weaver.[10] It is a very simple test that gives clear results.

The blood identification in mortars was analysed by the Kastle-Meyer's test. We used laboratory-prepared mortars and the results were satisfactory.

The identification of whale oil was achieved through thin layer chromatography.[11] In this case, we used the following substances:

Chloroform + methanol (extraction)
Chloroform (resuspension)
Chloroform + methanol + acetic acid + water (to develop the plates)
Phosphomobilidic acid (to reveal)

We used a small amount of a chloroformic solution of non-refined whale oil as a standard. During the course of the experiment we tried the solution at various degrees of concentration. As a preliminary result, we noted stains that indicated the presence of some oil components in the analysed mortars. As these stains may originate from different triglycerides, we must continue the analysis to complete identification. The following steps will be taken:

- to use different standards
- to modify the solvent polarity by using various amounts of the solvent or even different products as solvents
- hydrolysis and stearification of the samples, followed by GC or MS analysis
- identification of the different components by performing FTIR analysis.

Conclusion

It is very important to understand how our ancestors used building materials and why they did so. Sometimes an old technology lost for generations can still be reintroduced for use as an alternative way of building.

In Brazil, as in other developing countries, research in the field of building materials needs to be encouraged, in order to save money and to make use of locally available materials and skills.

Specifically, regarding the research on organic additives to lime, we still have much work to do. We must not only produce more conclusive results, but investigate other samples and additives as well. We hope to increase the interchange between different institutions, in order to become acquainted with work they are developing in the same field. Closer liaison will be fundamental for us, and will enable an exchange of greater knowledge, bibliographical information and perhaps human resources. It would then be easier to develop alternative construction, conservation and restoration techniques.

24. Lime stabilized lignite fly ash
BÜLENT BARADAN*

Coal-burning electric utilities annually produce millions of tonnes of fly ash as a waste by-product worldwide, and the environmentally acceptable disposal of this material is causing increasing concern. Past and recent research has established the potential of fly ash for use in a variety of construction applications, such as fills, concrete, pavements, grouts and others. However, there is still a need to find new uses and increase utilization, so less ash disposal will be needed.

Stabilized fly ash is a mixture of fly ash and lime, or fly ash and cement, compacted at optimum moisture content, and cured to form a product like soil-lime or soil cement. This monolithic material incorporates fly ash as a pozzolanic aggregate, while lime and cement serve as the activators to effect the stabilization reactions needed for strength development. Large volumes of fly ash can be utilized through this scheme because no other aggregate is added to the mix. Thus, stabilized fly ash appears to be a desirable alternative to the better known lime:fly ash:aggregate (LFA) and cement:fly ash:aggregate (CFA) mixtures, from the standpoint of mass utilization. Besides base and sub-base courses, potential also exists for using this material in the construction of liners for waste containment structures and canals, as well as seepage cut-off structures.

Limited field experiments, past studies and research[1,2,3,4,5,6] have shown that bituminous coal-based (class F) fly ash can be successfully stabilized with lime and/or cement. However, similar research on fly ashes derived

---

*Dokus Eylül University, Izmir, Turkey.

from sub-bituminous and lignite coals (mostly class C) has not been done to any appreciable degree, to our knowledge, and has not been well-documented in the literature. A laboratory study was thus undertaken to investigate and document the engineering and stabilization characteristics of lignite fly ashes. Two ashes produced in power stations located in western Turkey were selected for this project.[8] The materials were first tested in the unstabilized form to determine the various physical and chemical properties specified by ASTM standards to assess suitability for use as a pozzolana. This was followed by a series of compaction tests run on the ashes, and lime-ash stabilized mixes to establish the optimum moisture content and maximum dry density needed for stabilizing specimen fabrication. Unsoaked compressive strengths were then measured on specimens of the two ashes stabilized with varying percentages of lime and cured for different periods. The experimental programme was completed through studies involving soaked and vacuum-saturated durability, permeability and leachate characterization. The findings and conclusions of this research are presented herein in the same order as the experimental work.

Ash sampling and testing

The two fly ashes used in this study were Soma B, obtained from Soma B Power Plant in Manisa, Turkey and Yatagan acquired from the Yatagan Power Plant in Mugla, Turkey. They were collected from the dry hoppers in the power plants and transported to the laboratory for testing. The hydrated lime used in this study was manufactured by the Super Kireç, Mulga, Turkey, and was bought in paper sacks from local suppliers.

A variety of ASTM specification tests were initially performed on the fly ashes. These included specified gravity (ASTM D854), fineness, as established by the amount retained when wet-sieved on No. 200 and No. 325 sieves (ASTM D422), pozzolanic activity index with Portland cement and pozzolanic activity index with lime (ASTM C618), and lime pozzolan strength development (ASTM C593). A summary of the test results is presented in Table 1, along with the related ASTM specification criteria. Data on the chemical analyses of the ashes shown in the table were mostly provided by the Turkish Electric Company, which operates the power plants.

The specific gravity values shown in Table 1 indicate that the Yatagan ash is slightly heavier than Soma B which may be attributed to its higher Fe_2O_3 content. Also, Yatagan has a higher total amount of glassy components (SiO_2, Al_2O_3 and Fe_2O_3) than Soma B ash. Both ashes exhibit acceptable pozzolanic activity index with lime and cement. The lime-pozzolan strength development results for both ashes are within reasonable limits. Overall, Soma B ash shows much better pozzolanic reactivity. This is not surprising because CaO content of Soma B is significantly higher than that for Yatagan, suggesting that Soma B should have better cementitious quality.

Table 1. Properties of fly ashes

Property	Soma B	Yatagan	ASTM specifications Class C
Specific gravity	2.33	2.35	–
% Retained No. 30 sieve	0	0	ASTM C593 2.0 max
% Retained No. 200 sieve	18.24	8.4	ASTM C593 30.0 max
Fineness, % retained on No. 325 sieve	32.10	25.68	ASTM C618 34.0 max
Moisture content (%)	0.76	0.52	ASTM C618 3.0 max
Pozzolanic activity index with lime cement (%) (a)	95.0	76.0	ASTM C618 75 min
Pozzolanic activity index with lime (psi) (b)	925	841	ASTM C618 800 min
Water requirement, % of control	30	37	ASTM C618 105 max
Lime-pozzolan strength development (psi) (c)	1118	619	ASTM C593 600 min
Calcium oxide (Ca), %	22.93	12.74	–
Silicon dioxide (SiO_2), %	42.42	43.38	–
Aluminium oxide (Al_2O_3), %	27.56	27.71	–
Ferric oxide (Fe_2O_3), %	5.5	11.18	–
Sum of SiO_2, Al_2O_3 and Fe_2O_3	75.48	82.27	ASTM C618 50 min
Sulphur trioxide (SO_3)	1.83	1.90	ASTM C618 5.00
Magnesium oxide (MgO)	1.99	3.67	ASTM C618 5.0
Loss on ignition	0.76	0.52	ASTM C618 6.0 max

(a) Cured 1 day at 73°F plus 27 days at 100°F
(b) Cured 1 day at 73°F plus 6 days at 130°F
(c) Cured 7 days at 130°F

The loss of ignition values are comparable for both ashes, with Soma B being slightly higher. These results indicate that there are relatively small amounts of carbon and other combustible materials in the ashes. Overall, though quite different in properties, both ashes satisfy the ASTM specification criteria for use in cement and concrete as well as for lime-pozzolan stabilization.

Compaction

Compaction characteristics of mixtures of fly ash and lime were investigated by performing standard Proctor tests (ASTM D698) on materials using varying stabilizer contents. The maximum dry density (MDD) and optimum moisture content (OMC) were obtained on each mixture. It is apparent from the results that there are some differences between the compaction characteristics of the two fly ashes. Soma B produces higher maximum dry densities compared to Yatagan, although the difference between their specific gravities is not significant. However, Yatagan, being a finer material, yields higher optimum moisture contents because of its larger surface area.

The data also indicate that, for both ashes, increased stabilizer content results in increased MDD and decreased OMC after 5–10 per cent stabilizer content. These results can be attributed to the fineness and the low unit weight of lime. The moisture density relationships for individual mixtures were very straightforward to obtain, and the standard laboratory procedures posed no problems or anomalies.

Strength testing

The strength development in stabilized fly ash was investigated by compacting specimens in Proctor moulds at their OMC, extracting the specimens from the moulds and placing them in plastic bags, curing in the moist room at 73°F (23°C) and 100 per cent relative humidity, and testing them after specified curing periods for unconfined compressive strength in the unsoaked condition. The different ash stabilizer combinations and curing periods employed in this phase of the study are shown in Table 2, along with the test results. Both ashes were stabilized with 5, 10 and 15 per cent lime to assess the early setting effects of unconfined compressive strength at 3 days. Unconfirmed compressive strengths for unstabilized ashes (zero per cent lime) were also obtained to establish baseline values.

It can be observed from the results presented in Table 2 that increasing lime content may decrease the compressive strength following an increase up to some point. This is true of both ashes at 7 and 28 days curing periods. The trend is somewhat different at 3 days, showing a continual increase. It appears that lime is reacting well with the ashes in the short term.

As expected, there are considerable strength gains with extended curing periods for both ashes.

In comparing the strengths of the two ashes, it is seen that Soma B has developed higher strengths for lime stabilizations. This can be predicted from the pozzolanic reactivity test results shown in Table 1, for Soma B appears to have higher pozzolanic reactivity characteristics than Yatagan.

The results in Table 2 indicate that both ashes show quite satisfactory strength values with lime for all stabilized contents at 7 and 28 days for curing periods, with few exceptions. (The specified value for lime-stabilized sub-base courses is 100psi in AASHO standards). While both ashes exhibit

Table 2. Compressive strengths of stabilized ashes

Mixture	Compressive strength (psi)		
	3 days	7 days	28 days
Unstabilized Soma B	131	333	359
Soma B + 5% lime	438	700	790
Soma B + 10% lime	451	823	910
Soma B + 15% lime	482	736	841
Unstabilized Yatagan	57	70	83
Yatagan + 5% lime	201	320	350
Yatagan + 10% lime	254	574	613
Yatagan + 15% lime	320	517	811

relatively low strengths in the short term (3 days), acceptable values of strength can be obtained at higher stabilizer contents.

It is also observed that extending the curing period produces a significant effect on all mixtures. The longer the curing period, the higher the strengths.

Durability evaluation

Durability of lime-stabilized fly ash mixtures was evaluated by obtaining the residual strength, 'qr', of the cured specimens after subjecting them to two different exposure conditions, and comparing this value to the original (pre-exposure) strength, 'qo'. The first series of tests involved a 4-hour soaking in water. The second series of tests employed vacuum saturation with water which was done in accordance with ASTM C593.

Durability test results for the mixtures for both exposure conditions described above are presented in Table 3. The results are presented in terms of qr/qo, which is the ratio of the residual strength and the original strength. The original strength values used in computing these ratios are those given in Table 2 for the same mixtures. A qr/qo ratio of less than 1.0 indicates a strength decrease as a result of the exposure, while a qr/qo ratio of greater than 1.0 signifies a strength increase as a result of these exposures and testing procedures being adapted for durability evaluation. The particular criterion to be met by these parameters should be determined for the anticipated field and qr/qo ratios might indicate a need to evaluate critically the potential durability problems for the given case.

Several observations can be made from the data presented in Table 3. First, stabilized mixtures of Yatagan and Soma B ashes appear to have produced generally comparable durability. It is noted that the durability ratios increased in most cases with the longer curing periods for both ashes. This underscores the importance of obtaining sufficiently high strength in

Table 3. Strength retention upon soaking and vacuum saturation

Mixture no.	Vacuum sat. with water			4-hour water soaking		
	3 days q_r/q_o	7 days q_r/q_o	28 days q_r/q_o	3 days q_r/q_o	7 days q_r/q_o	28 days q_r/q_o
SL0	0.90	0.82	0.91	0.87	0.83	0.97
SL5	0.84	0.96	0.89	0.87	0.87	0.88
SL10	0.89	0.88	0.89	0.94	0.83	0.88
SL15	0.66	0.65	0.79	0.67	0.65	0.74
YL0	0.39	0.81	0.73	0.61	0.76	0.64
YL5	0.85	0.96	0.85	0.87	0.88	0.90
YL10	0.76	0.88	0.83	0.79	0.87	0.86
YL15	0.86	1.03	0.95	0.94	1.28	0.99

stabilized ash mixture prior to their exposure to possible detrimental service environments. Overall, it can be stated that increased stabilizer contents and extended curing periods enhance the durability of the stabilized ash mixtures. As part of the study, linear regression analysis of vacuum and 4-hour water saturation residual strength/original strength ratios were performed to compare the results of the two different durability evaluation techniques. Coefficients of correlation for both ashes for different curing periods varied from 0.857 to 0.902, indicating a very good correlation between the results of two testing techniques.

Permeability tests

An environmental concern has been the possibility of leaching of heavy metals from the fly ash into the underlying groundwater. The process of stabilizing metals in the fly ash and thereby preventing them from leaching can lead to a material which has a low permeability, resulting in reduced flow through the fly ash.

Selected mixtures of stabilized fly ash were tested to determine the effect of stabilization on permeability characteristics of the material. The samples of the mixtures to be tested were prepared by using standard Proctor compaction procedures, and allowed to moist cure for 28 days. A pressurized constant-head test was performed. The permeability tests were performed on the specimens using a rigid-wall set-up, without removing the moulds. The specimens were permeated until the permeability stabilized and until two pore volumes of flow had taken place.

The permeability of unstabilized Yatagan, compacted at optimum moisture content, was found to be 4.7 E–0.4cm/sec and for Soma B, the permeability was 1.7 E–0.4cm/sec. Both ashes showed a trend of decreasing permeability with the addition of stabilizer. At the 28-day curing periods and 10 per cent lime, Yatagan had permeability of 6.1 E–0.5cm/sec, Soma B showed a permeability of 4.8 E–0.5cm/sec. These values are higher than permeabilities measured on type F fly ashes.[7] Since none of these permeabilities met the often prescribed value of 1.0 E–0.7cm/sec for liners, additional tests should be performed with increasing stabilizer contents or addition of filler material like bentonite, if the material is intended to be used as a liner.

Leachate analysis

Not only does the stabilized fly ash contain the waste of the facility it is used to line; but, it must also contain any toxic elements inherent within its own matrix. The chief concern is the chance of certain constituents leaching into groundwater at concentrations determined to be hazardous to human health. These constituents include arsenic, barium, cadmium, chromium, lead, mercury, selenium and silver, among others. The US Environmental Protection Agency has classified these heavy metals as important contaminants and limited their existence in leachates.

Selected stabilized and unstabilized mixtures of fly ashes were subjected to the EP Toxicity testing procedures prescribed by EPA. The EP Toxicity test is meant to act as a surrogate for potential leaching from a municipal solid waste landfill. After pH adjustment of the mixtures the chemical analysis was performed by atomic absorption test. Concentration of all of the above constituents was determined, and the test results and the allowable maximum concentrations of these heavy metals specified by EPA are presented in Table 4. Atomic absorption test results indicated that the concentration of all heavy metals for both ashes were far below the stated EPA limits in unstabilized or stabilized form. Based on these results, it is safe to state that the ashes tested are not hazardous to groundwater.

Table 4. Results of EP toxicity tests

Element	Concentration (mg/l)				
	Unst. Soma B	Soma B + 10% lime	Unst. Yatagan	Yatagan + 10% lime	EPA limits
Arsenic	2.01	2.03	1.92	1.98	5.0
Barium	16.04	16.42	15.46	15.50	100.00
Cadmium	0.09	0.05	0.06	0.11	1.0
Chromium	0.44	0.01	0.00	0.00	5.0
Lead	0.35	0.38	0.32	0.39	5.0
Mercury	0.065	0.060	0.095	0.068	0.2
Selenium	0.18	0.22	0.20	0.22	1.0
Silver	0.00	0.00	0.00	0.00	5.0

Summary and conclusions

Engineering properties related to the stabilization of lignite coal (class C) fly ash were discussed in this paper. The findings of the studies described in the foregoing sections lead to the following general observations and conclusions:

1. Class C fly ash can be successfully stabilized with lime to produce a pozzolanic base course material, which does not require the addition of any aggregate.

2. The two fly ashes evaluated in this study exhibited adequate levels of pozzolanic activity and satisfied all the relevant ASTM specification criteria for class C pozzolans used in cement and concrete, and in lime-pozzolan stabilization. Notable differences were observed in the compaction, strength and durability characteristics of the stabilized fly ash mixtures.

3. The studies showed that adequate strength and durability levels can be achieved with stabilized fly ash by incorporating sufficient amounts of lime and allowing the mixture to cure for a sufficient period. Achieving adequate levels of strength prior to service exposure is important.

4. Increasing lime content caused an increase or decrease in strength depending on the stabilizer content and the length of curing. Extended curing, however, increased strength in all cases.

5. Durability ratios increased in most cases with the higher curing periods for both ashes. With few exceptions, increased stabilizer content and longer curing period enhanced durability.

6. The two exposure conditions, vacuum saturation and 4-hour water soaking, revealed durability ratios that are well correlated. It appears that the much simper test method of 4-hour water soaking may be used, in the absence of the vacuum apparatus.

7. The permeability test results indicate that lime stabilization can reduce the permeabilities of both ashes. However, the permeabilities were higher than expected. Addition of fillers like bentonite, or raising the stabilizer percentage, must be considered for lowering permeability.

8. Leaching tests by the atomic absorption method revealed that unstabilized or stabilized fly ash mixtures are environmentally acceptable.

9. The initiation of field trials and demonstration projects using stabilized fly ash, lime/fly ash mixes in particular, should be encouraged to establish performance records, which can help the highway industry determine the advantages of utilizing this material.

25. The use of alternative binders in Rwanda: a case study

THEO SCHILDERMAN*

With around 6.5 million inhabitants on 26 000 km², Rwanda is the most densely populated African country. Around 90 per cent of its population is rural and living from agriculture, on plots that are already too small to provide a decent living, but the country possesses very little alternative resources that can be developed.

Traditionally, the Rwandese live scattered over the many hills of the country, each family on its own plot. The majority of dwellings are constructed with mud-and-pole walls, but soil blocks are becoming increasingly popular as a walling material. Roofs are covered by thatch, tiles or corrugated iron. Floors are most commonly of earth, and sometimes paved.

The weather in Rwanda is often cold and damp. The climate has a negative impact on dwellings built of soil, which often experience problems such as: rising water or damp in the foundations and basements, erosion of walls and basements by rain, cracks in walls harbouring vermin, or damp floors. These conditions can be dealt with, but many potential improvements rely on a binder. Until the mid-1980s, ordinary Portland cement (OPC) used to be imported; it was expensive, and sometimes scarce. Moreover, OPC mortars are relatively rigid and impermeable and therefore less suitable for the improvement of earth-based wall elements. These were important reasons for the development of an alternative, local binder, as described in this case study.

Project idea and objectives

The PPCT (Pozzolana-Lime-Peat project) was initiated by the Belgium-based NGO, COOPIBO (Development Co-operation of the International Building Companions), following research on local building materials implemented in another project during the mid-1970s. In those days, research into alternative binders was aimed at import substitution, since an average 20 000 tonnes of OPC were imported over the years 1975–77, and at the development of a cheaper alternative to OPC. Initially, the work focused on the promotion of lime production and the partial substitution of OPC by lime. Subsequently, the emphasis shifted towards the upgrading of lime into a lime-pozzolana. Rwanda possesses extensive deposits of volcanic pozzolanas and limestone, mainly located in the NW and SW of the country.

*ITDG, UK.

After some initial studies, casting some doubt on the reactivity of the pozzolanas tested, a pilot project was nevertheless established near the town of Ruhengeri, in NW Rwanda, on a site possessing minor limestone deposits, water, a peat marsh and some forest. This decision was taken in rather a hurry, under pressure by the Rwandese authorities, who wanted fast results. No feasibility study was undertaken at that stage, and it would later become clear that a lot more surveys and research would have to be undertaken to get production off the ground.

COOPIBO signed a convention with the Rwandese authorities, then represented by the Ministry of Planning, in 1979. Funding throughout the project was mainly provided by the Belgian Ministry of Co-operation, with increasing problems as years went on.

The project was designed in line with the development philosophy of COOPIBO, emphasizing an alternative development model with priority for rural areas and their population, who would determine and manage its own development. The production of cheap binders and the creation of employment were mere steps on the way towards that overall broader goal. The project document mentions the following objectives:[5]

- to establish a company that would provide employment to landless peasants, which would be managed by the labourers in the form of a co-operative
- to produce building materials from local resources, for import substitution: masonry cement, lime and peat
- to create employment in a region where half the population did not possess enough land for subsistence
- to produce an affordable binder of satisfactory quality, and to study its application and dissemination
- to advise the rural population on housing improvement
- to introduce peat as a residential and industrial fuel
- to support popular development initiatives.

Towards production of alternative binders

It proved not to be easy to develop this company, under such diverging objectives, and often by trial and error, since experience with similar materials elsewhere was very scarce. In the first years of the project, technical research and development received inadequate attention, and emphasis was placed on the social aspects of the work, including participation, organization, literacy and other forms of training. Thus, overall progress on the technical side remained slow initially, and surprises were inevitable.

- The limestone deposit on site was not worthwhile: small, impure and difficult to exploit. Subsequently, two other small deposits, at Rwaza

(12km) and Mpenge (5km) have been explored, opened up and are currently exploited.
- A first brief survey showed that volcanic pozzolanas deposited in the form of small grains (lapillis) at Mutura, 40km along the main road to Gisenyi, were the most reactive. The deposit is rather far away, but currently along the tarmac road. Subsequently surveys of alternative deposits have not come up with more reactive pozzolanas, and access to those sites was often difficult.
- The peat marsh proved difficult to drain. This was originally achieved via a hand-dug channel, but later on additional pumping was required. Peat is used for drying the pozzolanas, which arrive at 15–20 per cent humidity, to less than 1 per cent, before grinding. It was also intended to be gasified for lime burning, but without success. Lime burning with a mixed feed of peat lumps did not work, but caused rather too many kiln blockages, and was therefore abandoned in favour of eucalyptus wood as fuel. The project subsequently planted most of its site with eucalyptus.
- It was originally intended to produce a binder on the basis of ground pozzolana and lime only. Accelerated curing tests, at higher temperatures, had given some hope in that direction, but these were not borne out in practice. The pozzolanas of NW Rwanda are insufficiently reactive to make a satisfactory lime-pozzolana, and an accelerator needed to be found if the binder was to be used in regular construction work. After a lot of research, e.g. including soda ash from Kenya, it was concluded that OPC would be the easiest to obtain and to use as a kind of accelerator. This makes the final product a blended cement.
- It was also hard to design a factory with the limited experience of project staff, and the absence of similar models elsewhere. A fair amount of the technology had to be developed locally, e.g. drying kilns for pozzolanas, a lime kiln working on peat, and a motorized lime sieve. It was unavoidable in the end to import some of the machinery, including a vibrating ball mill and a mixer. The mechanization and the scale of the production process in its turn required more professional management, and moved the company away from the original co-operative idea.

Production of a binder finally took off in late 1983; the product, which can probably be classified as a masonry cement, reached compressive strengths of $5N/mm^2$ after 7 days, and $10N/mm^2$ after 28 days. This chapter does not cover the research and development of the final product, nor the quality of its components, which are dealt with elsewhere. It is sufficient to mention that, in its ultimate composition, the cement included the following components (by weight):

- volcanic pozzolanas (66.2 per cent), from Mutura, at 40km

- lime (12.5 per cent), from Rwaza at 12km, OPC (20 per cent), of quality P 425, produced in SW Rwanda since 1984, and supplied via a middleman in Gisenyi, at 60km
- gypsum (1.25 per cent), imported from Kenya
- trietanolamine (0.04 per cent), an imported grinding aid, helping to clean the steel balls in the mill, and thereby resulting in increased fineness and strength.

The production process included drying the pozzolanas on open platforms moving vertically through a shaft kiln heated with peat. Limestone, broken down to fist-size lumps, is fired with eucalyptus wood of approximately 40cm length, in a continuous vertical shaft kiln, of an internal diameter of approximately 1.20 × 1.80m, by a height of 9m, with a capacity of at least 250 tonnes of slaked lime per month. Components are weighed on simple balances, mixed, and stored; this mixture is subsequently transported by wheelbarrow to a hopper above the vibrating ball mill, of a capacity of 6 tonnes/day, at a fineness of 4000–5000cm^2/g (Blaine). The final product is packed in paper bags, closed by sewing.

Production and sales then developed as shown in Table 1 in subsequent years (in tonnes/year).

It should be noted that, apart from the lime appearing in the table, a substantial quantity was produced for internal use in producing masonry cement. Masonry cement production in 1986 reached a peak of just over 2050 tonnes, its installed capacity, but subsequently declined.

The use of masonry cement

One of the main objectives of the PPCT was to develop a binder that could be useful and affordable to lower-income groups, particularly for shelter purposes. The project recognized that an effort was needed to establish and promote the proper use of the binder. This was attempted through a major research programme, during 1984–86, composed of four phases:

1. a survey of the use of binders traditionally used in informal housing: mortar composition, techniques of application, cure, quality assessment, etc.

Table 1.

Year	Lime		Masonry cement	
	Production	Sales	Production	Sales
1984	0	?	340	300
1985	80	140	1225	1175
1986	475	101	2050	2050
1987	760	930	1227	1170

2. application of the most promising compositions and techniques on a range of test elements (plasters, pavements) but substituting the traditional binders with masonry cement
3. application and monitoring of the most satisfactory masonry cement mortars on a number of pilot sites outside the project
4. elaboration of characteristics and guidelines for the use of the masonry cement.

Since funds were limited, this research programme had to concentrate on the technical aspects of the use of binders, and was therefore unable to touch upon the socio-cultural aspects, including preferences, traditions, and risks.

The use of binders in informal housing of NW Rwanda[1]

A survey was made of the actual use of binders, such as lime and Portland cement, in four municipalities of NW Rwanda. These municipalities were chosen on the basis of their location along major road axes (hence good supplies) and differences in local resources such as soils and sands. In each municipality, three sectors were identified with a certain concentration of houses using binders for plasters and pavements, and a maximum of six houses were surveyed per sector. After initial contacts with municipal and sector leaders, the actual surveys were conducted with the masons who had executed the jobs and the house owners.

Important differences were noted between municipalities, depending, for example, on traditions and locally available materials. Otherwise, differing results could have been caused by the use of different sands, masons, locations, etc. Binders are primarily used for plastering walls, and subsequently for pavements. They are used far less for masonry work, foundations and basements.

In this chapter, only the use of binders for plastering walls is described, as an example of the research and resulting determination of masonry cement compositions and methods of use. All mortar mixes mentioned are as measured on site, that is by volumetric proportions of binder to sand:

a. Plasters on brick walls (52 cases: 33 internal + 19 external).
- 77 per cent used OPC; most common mixes are 1:5à7; best results are obtained with mixes in the range 1:4à6.
- 18 per cent used a lime-cement mortar, most commonly 1:4à5. (Note: although the lime:cement proportion can occasionally vary considerably, the cement content is normally quite small, around one fourth to a fifth of the volume of lime). Some of these plasters show small fissures.

b. Plasters on soil blocks (79 cases: 43 internal + 36 external).

- 37 per cent used OPC; most common mixes are 1:5à8; often a first layer of soil plaster is applied; the quality of these plasters varies considerably.
- 26 per cent used a lime-cement mortar (see above note), most often of a 1:4 mix, and directly applied to the blocks. A 1:4 mix seems to be the upper limit to achieve good quality. In this case, it is noticed that it is less critical to apply a first layer of soil plaster, because the largely lime-based mortar adheres better to the blocks than an OPC mortar.
- 37 per cent used other plasters; this occurred particularly in one municipality, where a mortar made out of a local soil (kigagwa) and sand often produced good plasters, particularly when finished with a limewash (internally) or sprayed OPC plaster (tyrolean plaster externally).

c. Plasters on mud-and-pole walls (36 cases: 18 internal + 18 external).
- 56 per cent used a lime-cement mortar (see above note), with a most common mix of 1:4. This mortar is preferred because of its superior adherence. Often, a first layer of soil is applied or a lime-sand mix, and this produces better results than plastering directly on the mud-and-pole. Mixes of 1:3à4 with a finishing lime-cement wash provide good results.
- 24 per cent used an OPC mortar, on a base layer of soil. The usual mix is 1:6; richer mixes tend to fissure more, because of the rigidity of the cement plaster.

d. Plasters on stone walls (21 cases: 8 internal + 13 external).
- 55 per cent used a lime-cement mortar (see note above), usually 1:4; two-thirds of those plasters were good quality.
- 45 per cent used OPC mortars, generally 1:6, which were mostly of good quality as well.

It is interesting to note that, traditionally, a largely lime-based plaster is used more frequently on soil walls (categories b and c) than on OPC plaster, and they often produce better results. The OPC plasters do better on the more rigid brick and stone walls.

Tests on plasters with masonry cement

Following the field survey, a series of test elements were then executed, under strictly controlled conditions, on the site of the project; plasters were selected with similar compositions to the ones that had behaved well in the informal sector, but this time substituting lime and OPC with the new masonry cement. Two different wall materials were chosen: fired clay bricks and soil blocks, which were thought to adequately represent the categories of wall materials used in practice. All test walls had been built a considerable time (in most cases years) before the plaster was applied. Once executed, plasters were monitored for at least six months, with the results as shown in Tables 2 and 3.

Table 2. Plasters on brick walls

Case	Subsequent layers of plaster	Results
1	1:3.5 + 1:3.5 + cement wash	Very good
2	1:6 + 1:6 + limewash	Good, but small cracks
3	1:6 + 1:6	Good
4	1:4 + 1:6	Good
5	1:6 + 1:4	Good
6	1:5 + 1:7	Good, but small cracks
7	1:7 + 1:5	Good, but small cracks

Table 3. Plaster soil blocks

Case	Subsequent layers of plaster	Results
1	1:5 + 1:7 + limewash	Good, but some cracks
2	soil + 1:5 + 1:6	Good, no fissuring and good adherence
3	1 soil: 2 sand + 1:5 + 1:6	Very good
4	1 soil: 2 sand + 1:7.5 + 1:7.5	Weak, medium quality
5	1 kigagwa: 1.5 sand + 1 kigagwa: 2 sand + 1:8	Weak, many cracks
6	soil + 1:8 + 1:8	Weak, medium quality
7	1:7 + 1:3 sprayed plaster	Weak, medium quality
8	1:6 + 1:2 sprayed plaster	Good
9	1 kigagwa: 2 sand + 1 kigagwa: 1:5 sand + 1:3 sprayed plaster	Medium quality
10	1:7 + 1:3 sprayed plaster + 1:3 sprayed plaster	Good

Plasters in the 1:4à6 range of mixes on brick walls did well, and so did the ones in the 1:5à6 range on soil block walls, provided a base layer is applied preferably of a soil-sand mix. Weaker mixes cannot be recommended.

External dissemination

The next stage of research dealt with the application of the above lessons to building sites around Ruhengeri, where control would be less strict. The project provided advice on the proper application of the masonry cement, but a number of variables are influencing the outcome, e.g. the skills of the mason, the quality of the sand and water used, and curing practices.

The dissemination took place on house building sites, at the request of the owner, and was partly subsidized. The programme lasted five months, and produced good results for plasters, but less so for pavements. Eighteen cases of plasters were tested, of which the ten shown in Table 4 provide an adequate picture.

Results were almost as good as under the more controlled conditions at the project site, and convinced the researchers that adequate plasters could be made out of 1:4à6 mixes of masonry cement mortar.

Table 4.

Case	Wall base	Subsequent layers of plaster	Results
1	Soil block	1:5 + 1:5	Good, no cracks, good adherence
2	Soil block	1 soil + 1:5 + 1:5	Medium quality, plaster too thick and peeling off
3	Soil block	1 soil: 2 sand + 1:5 + 1:6	Very good
4	Brick	1:5 + 1:6	Good, but few small cracks
5	Mud and pole	1 soil: 2 sand + lime: sand + 1:5 + 1:4	Good
6	Mud and pole	1 soil: 2 sand + 1:4 + 1:4	Good
7	Mud and pole	1 soil: 2 sand + lime: sand + 1:5 + 1:3	Good
8	Mud and pole	1 soil: 2 sand + 1:5 + 1:5	Medium quality, fissuring
9	Mud and pole	1 soil: 2 sand + 1:5 + 1:7	Good
10	Mud and pole	1 soil: 2 sand + 1:6 + 1:2 sprayed plaster	Medium quality, lacks adherence

Guidelines for the use of masonry cement

The previous research programme produced clear indications with respect to the proper composition and use of masonry cement mortars. These were published in a manual[2] providing information about:

- the general characteristics of the product (strength, fineness, hardening)
- general conditions of use (choice and quality of granulates and water, surface preparation, curing)
- conditions of use for specific purposes (masonry, pointing, various plasters, pavements, light concretes)
- advisable mixes, in the form of Table 5.

Table 5.

Element	Mix by weight (kg) per m³ sand	Mix by volume cement:sand
Masonry mortars	175–250	1:5à7
Pointing of walls	175–250	1:5à7
Pointing of basements	300–400	1:3à5
Plaster on bricks, stone	200–300	1:4à6
Plaster on soil walls	200–400	1:3à6
Sprayed plaster finish	600	1:2
Pavements and drains	250–400	1:3à5
Water reservoirs	600	1:2
Light concrete	200–250	1:5à6

Marketing developments

Table 1 shows the evolution of masonry cement production and sales over the years 1984–87. Production reached a peak in 1986 of just over 2000 tonnes, which was then the installed capacity, but subsequently declined.

Market studies, done over the years for various types of cement (depending on the stage of research), had always been quite optimistic about the potential market. The latest study, by a Belgian expert, in January 1987, concluded that it would be economically feasible to expand the factory to a capacity of 6000 tonnes/year, and that the effective demand for masonry cement would rise from 12 500 tonnes in 1986 to 15 500 tonnes in 1990. The latter was due to the fact that the OPC factory, in SW Rwanda, had reached its capacity, and shortages of OPC would appear. A subsequent study, by a Belgian mission of cement experts, concluded that a capacity of 3000 tonnes/year would be the minimal economic size of the factory, and that on the condition that the OPC factory would supply clinker, instead of the then used Portland cement in powder form.

Practice proved to be quite different, which shows how unreliable market studies of innovative materials can be. The actual sales declined in 1987, and the masonry cement production ceased altogether during 1988. There were a number of reasons for this development. The continuing decrease in price of OPC, which diminished from 30 000 FRW/tonne (ex factory) when it first came onto the market in 1984, to 21 500 FRW/tonne in 1988. Over the same period, the price of masonry cement diminished as well, from 17 000 FRW/tonne (ex factory) to 15 000 FRW/tonne, but from being at 57 per cent of the price of OPC, it had now gone up to 70 per cent. Although prices on site differ, due to transport costs and profits, masonry cement had definitely lost its advantage in most areas of the country. In theory, one needs richer mortars with masonry cement to achieve similar results to Portland cement mortars; one may have to add 50 per cent more cement to the mix. With the above prices, there is no real advantage. In practice, builders often use very diluted OPC mortars, with which the masonry cement can hardly compete. Once the public noticed there was no real financial gain in using masonry cement, it lost interest in the product, and returned to OPC.

There is also a risk involved in using masonry cement. The margin of error, e.g. caused by improper application of curing, is much smaller with masonry cement than it is with OPC, which is hugely overqualified for most of its uses. The public does not like to take risks with an innovative product, particularly if there is no financial gain in it. It would take a considerable effort of dissemination, training and advice, which would be expensive, to guarantee that such risks are avoided in the use of masonry cement, which would add to this cost or would have to be met through external aid.

Other factors included:

- Affordability to lower income groups. It is questionable whether, at 750 FRW/50kg/bag, masonry cement is really affordable by the largest potential market. Where official minimum wages stood at only 100 FRW/day, and GNP per capita at around 25 000 FRW per year, the cement seems to have overshot the market that matters most.
- Absence of a substantial market in the formal sector. The PPCT managed to sell to a few big clients, but did not achieve a decent market share. Reliance on a few big clients made the company vulnerable to fluctuations in building activity.
- Politics. There was in the end a lack of political will to promote the product. Although the Ministry of Works had accepted the Guidelines for the use of the masonry cement as a kind of standard, and allowed it to be used on public sites, it actually never prescribed the product in its contracts, which could have made a huge difference by its example. The Minister of Industry, who was the main partner of COOPIBO in later years, and quite interested in the project in the mid-80s, lost interest towards the end, partly caused by a change at the top. When the OPC factory refused to sell clinker to the PPCT, which would have been essential to expand production and lower the price, the Ministry did not interfere. Obviously, it did not wish to upset one of the largest industries of the country, source of much tax income, and backed by the Chinese.

The current PPCT[6]

When, in 1988, it was decided to cease production of masonry cement, the workers in the factory requested to continue the production of lime, in the form of a co-operative. This proposal was not directly acceptable to the authorities, to whom the project had been handed over in the meantime. Alternative forms of enterprise, such as joint ventures, have been investigated in the meantime, but currently the co-operative option still seems the most feasible. Whilst the organizational structure is being sorted out, production of lime continues, and reached 3175 tonnes in 1990, which is at the limit of the kiln's capacity. Three qualities of lime are being produced:

- lime from Rwaza limestone, sieved at 0.5mm, which exceeds 80 per cent $Ca(OH)_2$, and is sold for water purification, sugar refining, leather processing and whitewash
- lime from Mpenge limestone, sieved at 0.5mm, which exceeds 70 per cent lime $Ca(OH)_2$, sold for non-industrial purposes
- lime from milled sieve residues, reaching about 70 per cent $Ca(OH)_2$, sold for agricultural purposes, building works and cattle feed.

There has been a large increase in demand for lime, particularly from the agricultural sector, and the project has in fact bought up and reprocessed 1032 tonnes of lime produced by the artisanal sector, to satisfy the 1990 demand, which divides as follows:

- 2712 tonnes (63.5 per cent) to agriculture
- 770 tonnes (18.0 per cent) to water purification
- 778 tonnes (18.5 per cent) to builders, trades, others.

In addition, the PPCT produced and sold minor quantities of milled limestone, limestone granulate, the masonry cement, the latter to an old client with an ongoing building programme. It made a profit, but this would have been insufficient to pay back all the investments, had the co-operative been required to do so. It employed 38 permanent staff and 90 part-timers at the factory, and a further 118, later increased to 188 workers in the quarries. It is obvious that there is scope to expand lime production, by building a second kiln; that would make the enterprise more profitable. The investment required is not enormous (around 5 million FRW), but the company is hampered by the current absence of a legal structure. There are also concerns about supplies of limestone running low, and one would have to explore other deposits. Timber is also a bottleneck; the project's own plantation had to be cut down in 1990 for military reasons, following invasions from nearby Uganda, and one may have to revert to peat at some stage.

Evaluation

The project originally aimed to establish a socially appropriate technology, based on small-scale production using mainly local resources and little energy, and relying on worker participation and a decentralized organizational structure. It has only achieved limited success: the masonry cement failed, but production of lime increased considerably in quality and quantity. The production is still small-scale, and some innovative technologies have been developed. There is also a degree of imported mechanization, and considerable energy consumption. Worker participation is limited. A comparison of the original objectives with the final outcome shows this in somewhat more detail in Table 6.

Looking back, the project failed most in its developmental aims: it has not become a real focus of rural development, although it may have contributed indirectly through wages to hundreds of people and the provision of lime. The initial objectives were probably far too ambitious for any NGO with no particular experience in this area (and experience elsewhere being in short supply). There was also no clear priority and maybe even some contradiction between objectives, which subsequently led to changes in strategies over the years. Where the initial focus was very much on the

Table 6.

Objectives	Achievements
To establish a company providing employment to the landless, with a co-operative structure.	A fair number of jobs were created, but participation is limited and the co-operative still remains a request.
To produce local materials as import substitution: masonry cement, lime and peat.	Lime is the only success story but peat still has potential.
To create employment.	Equivalent of over 200 man years employed.
To produce an affordable binder of good quality, and study its application and dissemination.	An adequate binder was produced and disseminated, but lost in competition with OPC, and was too expensive.
To provide advice on housing improvement to the rural population.	Only achieved on a very limited scale, and never formalized as activity.
To introduce peat as a residential and industrial fuel.	Peat was exploited, but is not currently used in the factory; household level.
To support popular development initiatives.	There were insufficient profits and too many other problems, to even start considering such initiatives.

area of social development and organization, and somewhat neglected research, thereby failing to produce a binder, the emphasis shifted in the end very much towards the technical and business aspects; that certainly got production underway, but was perhaps out of touch with the real market.

As a conclusion, the project should have been better prepared, e.g. using a better planned framework.

26. Standards for building limes

MICHAEL WINGATE*

The user of any product needs some assurance that it is appropriate for the intended use. If there are no agreed standards for the product he must rely on his confidence in the reputation of the producer, tempered by his own judgement based on experience.

When trade is only within a small locality a purchaser knows very well the types of product on offer and can probably discern one from another

---

*Chartered Architect, UK.

with his own senses by appearance, feel or even by smell. Thus a builder in the town of Rugby in England during the last century would have known that his local Blue Lias lime (with its buff colour) was excellent for masonry, giving a good early set and a strong bond. For plasterwork he might well have brought in White lime from elsewhere; this would have been a weaker set, but responded better under the plasterer's floats and trowels. He was *confident* of success with those materials.

Quality

Product quality might vary, but if it falls, the producer – the first to know that things are amiss – withholds the poor product to protect his reputation. It would be comforting to believe that!

When poor product quality was a common problem it was sometimes necessary to introduce laws to govern the trade. For example, in Ancient Rome lime plastered walls were often richly painted. Now one possible defect of lime can cause damage to the surface of a lime plaster and this may occur many months after the surface has been prepared and painted. As the damage can be avoided by using well matured lime a law was passed to ensure that lime for finishing plaster was stored as lime putty for three years; this was an early example of a building standard.

Such intervention by law was rare and a more common control was through trade associations or guilds of craftsmen which demanded good work of their members even if no formal standards were set. But reputation remains very important, as does good local knowledge and a general alertness to quality. Where these are found formal standards are hardly necessary.

New products, competition and widespread trading

The confidence of that builder in Rugby was rooted in a stable system. But this confidence can be undermined in conditions of change where new materials are developed or bought in from a distance, or where buildings are specified by an independent designer rather than by the builder himself.

In 1795 James Parker took out a patent in England for a 'Roman cement'; it was far stronger than any limes ever used and was a commercial success. It was made from a natural limestone with a very high clay content.

A generation later patents appeared for artificial or 'Portland' cements which were made from blends of chalk and clay. This blending made careful control possible. Competition between 'Portland' and 'Roman' cements was ferocious, including public cement contests.[1] Portland cement was the undoubted winner and perhaps its most conclusive advantage (in the early years) was that the material was produced to a standard specification, providing a reliable prediction of the strength which could be obtained. A serious structural failure at Euston Station in London was blamed on the Roman cement; the problem was not so much that the Roman cements were weak, but that their strength could not be confidently predicted by the designer.

In this example a working standard was produced by an association of manufacturers as a vigorous marketing device, but it became increasingly clear by the beginning of the 20th century that agreed National Standards could not only be the basis of confident design but also of a fair basis for competition. With the growth of international trade there is now a widespread demand for International Standards.

Preparing standards

Agreement at committees

National standards are drafted through the agreement of committees which include representatives of the producers and users of a product together with scientific, legal and administrative advisers. Inevitably each member strives to achieve the best result for his own organization and the agreements reached, although setting a baseline for confident use of a product, are not always wholly beneficial. They can be exclusive.

Poor standards may damage a lime industry

In the UK, the first standards for building lime were drafted in such a way as to reduce the role of lime in building to little more than an additive to improve cement mortars and gypsum plasters. The underlying problem was that the great diversity of limes then produced was not matched by a wide range of product descriptions and specifications which could encourage the best use of each lime. The building industry was conditioned to think of 'building lime' rather than 'building limes'.

The problems had been foreseen and discussed in a special report 'Lime and Lime Mortars' published by the Department of Scientific and Industrial Research (Building Research) in 1927.[2] This described the materials and their limitations and even proposed tentative standards, but all to no avail. The diversity of UK limes was lost and nearly all of the lime now produced for building is of one type.

Variations between national standards

Other countries achieved very different standards. The French Standard for natural hydraulic limes (NF P 15–310), for example, encourages the admixture of hydraulic lime and Grappiers cement (a natural cement) to achieve predictable strength standards. In other European countries, the standards encourage the production of lime-pozzolan masonry cements. In the US, the standards encourage premium grades of hydrated lime for finishing plasters. Such diversity must have arisen from the strength of the various producers and the range of the discussions at the committees.

Shared standards

The national standards for lime in the European and North American countries naturally reflect the strengths and scale of the lime producers in those countries. To some extent they are written to protect the interests of the relatively large-scale producers. Their adoption in developing countries would not necessarily encourage the existing small-scale producers and may, as in the UK, reduce the range of limes available. This would diminish the potential for the use of lime as the basis of good construction.

The preparation of appropriate national standards is time-consuming and costly. Differences between standards in neighbouring countries provide a barrier to free trade and there is every reason to look for common standards. But such standards must not exclude small-scale producers by requiring excessively costly test procedures, nor must they exclude producers of less than 'ideal' limes.

Identification of classes of lime

A good description of a lime must make clear three things:

- the TYPE of the carbonate from which the lime was made
- the FORM in which the lime is traded, and
- the QUALITY of the product, showing how good it is of its kind.

New standards must identify possible classes of limes which may have widely differing properties and be appropriate for a variety of uses.

Types of lime

One such distinction is the extent to which the lime has impurities; these may be inert or active.

Lean limes have inert impurities, as if sand had been added to a pure lime. This must reduce the value of a lime, but need not rule it out of consideration for many uses. The normal terms for this distinction are:

- fat limes (limey limes!)
- lean limes

and the lean-ness would be described by the proportion of impurity. The boundary must be drawn sensitively since everyone prefers a fat lime, but no limes are entirely free of impurities.

Hydraulic limes are those with impurities having an extremely fine particle size (as clay). They may be activated during the limeburning and have good setting properties, even under water. In general terms the range is this:

- pure limes ('air limes')
- somewhat hydraulic limes ('feebly' and 'moderately' hydraulic)
- very hydraulic limes ('eminently' hydraulic)
- natural cements (which need to be ground down)
- pozzolans (which contain negligible lime; lime must be added).

Great care is needed to establish helpful boundaries between the classes. Typical characteristics which have been used to distinguish hydraulic limes are the strength of mortar cubes, the speed of setting and the chemical composition; these properties are all interrelated.

Magnesian limes. Whilst all limes contain calcium oxide (or hydroxide) some also contain magnesium oxide (or hydroxide). For building purposes the magnesium compounds behave in ways very similar to the calcium compounds but there are some important differences relating to calcination temperatures, reactivity and slaking performance.[3] One possible classification of this range is:

- high-calcium limes (with negligible MgO)
- calcium limes (low MgO content)
- magnesian limes (significant MgO)
- high magnesian limes
- dolomitic lime (from pure dolomite)

but, again, boundaries must be set.

Magnesian limes may also be more or less hydraulic and more or less lean, so the classes will be used in combination.

Form of limes

All limes, of course, are manufactured as quicklime and in the past it was normal for lime to be traded as quicklime for slaking by the builder. But in the last fifty years it has become increasingly normal for lime to be traded in the slaked forms and particularly as the dry hydrate. For some uses lime is also traded in the putty form.

Quicklime (calcium oxide) is likely to be described in terms of the particle size using terms such as:

- lumps
- pebbles
- granular
- pulverized

Hydrated limes (forms of calcium hydroxide) may be dry or combined with excess water. The range is:

- dry hydrate
- lime putty
- milk of lime
- limewater

There is no advantage in trading in milk of lime as it will settle out to form lime putty and limewater. Limewater is unlikely to be traded as it has very limited use and is easily produced on site.

Quality of lime

The quality can be described both in terms of good attributes and of relative freedom from defects.

Reactivity of a lime. The heat regime in a lime kiln will affect the quality of a lime and its performance in use. The most reactive limes (produced at the lowest temperatures) have a high surface area which allows them to enter most readily into chemical reactions. Reactivity is most readily seen by the slaking performance of the quicklime, but also affects the activation of hydraulic setting properties and the setting of pure limes by reaction with carbon dioxide ('carbonation'). Limes of any class may be:

- lightly burnt (and highly reactive)
- hard burnt (and of only moderate reactivity)
- overburnt (of dangerously low reactivity)
- or possibly even dead burnt (almost inert).

The type of lime also has a significant effect on its reactivity.

Water retention is a virtuous property. As putties age (in excess water) they continue to absorb water into their physical mass and they hold onto this water tenaciously in use. This maturation also allows extra time for slaking of any overburnt particles. The age of putties is very significant if they are used in plastering, but can a claimed age be verified?

Overburnt lime, having a low reactivity, can cause problems. When overburnt quicklime is slaked some particles may remain as quicklime and eventually slake at a much later date, causing damage to finished work. It was this that the Romans were overcoming with their legislation for lime putties to be well matured. Careful sifting can remove the relatively large overburnt particles and avoid the defect of *pitting and popping*, but the late hydration of very fine overburnt material within a plaster can cause a general expansion known as *unsoundness*. A standard must ensure freedom from these defects, or at least restrict their severity.

Underburning is also a defect. When the carbonate is converted into quicklime in the kiln, the larger lumps will usually contain a core of unconverted material. This is harmless for most purposes (though it drags in a fine plaster finish), but serves to dilute, and thus devalue, the useful hydrate product. Sifting or air separation can remove the underburnt particles.

Standards may show a maximum permissible proportion of carbonate in the hydrate or in the quicklime.

Premature carbonation may affect a badly stored dry hydrate. The powder may combine with carbon dioxide from the air which forms a shell of carbonate around the hydroxide particles. This acts as a shield, preventing the lime from forming a good putty with water. This is rarely defined and is difficult to quantify, but may be important.

Air slaking is a defect in quicklimes. Where quicklime is needed for building purposes, relatively large lumps and pebbles are less likely to degenerate by *air slaking* in transit and storage. Dense lumps (from limestone rather than chalk) are still less likely to degenerate. When a quicklime air slakes it combines with moisture and carbon dioxide from the air and forms a powdery skin of hydroxides and carbonates until eventually the lumps collapse. We think of this as a serious defect justifying rejection of the lime, but for some special uses this has, in the past, been highly valued.[4]

Testing and certification

This is the heart of the matter. It is very easy to construct tight standards which rely on expensive test procedures which might be quite beyond the reach of all but the largest producers.

Difficult tests

An example of this is the problem of ascertaining the *plasticity* of a lime putty. This property is particularly important for plasterwork, as it shows how readily the plaster will spread; any plasterer will know what is acceptable and what is not simply by the feel of the material. But to test this in the laboratory requires elaborate machinery.

Two-stage testing

A better way forward might be to attempt to establish low-cost field tests to indicate the qualities of limes. Such tests might have very low accuracy, even as poor as 10 per cent. They could be backed up, in case of any dispute, with more elaborate tests. If possible these back-up tests ought to be simple enough to be executed in a secondary school laboratory.

Many of the properties of limes are interrelated, so if a field test cannot be devised for one property it may be possible to test for a related property. As far as possible the tested properties should be those which a practical builder would perceive. The 'plasticimeter' is removed from the realities of building, but a builder knows that a 'fat' lime will carry more sand than a leaner lime. It is the plasticity of the lime which enables it to spread more finely around the sand grains to bind them together, but could the field test concentrate on the carrying capacity? A good fat putty might carry 4.5 times its own volume of a certain sand whereas a leaner lime might carry only 2.2 times its volume. Guidance would need to be given on the degree of mixing (quite intense) and on the limit of cohesiveness and the choice of sand.

The twin defects of 'pitting and popping' are caused by poor slaking and overburnt particles. A simple field test (after SE Young) could be to spread a thin layer of lime putty, about 1.5mm thick, on a sheet of glass and examine it against a strong light. If dark spots are noticed, there is a possibility of pitting developing later on.

Tests for the nature of limes

These are likely to be rather slower to perform, but a quick indication can be gained by dissolving the *carbonate* in cold dilute hydrochloric acid. Calcium carbonate dissolves quite readily whilst magnesium carbonate will only dissolve if the acid is heated. Thus a cloudy residue suggests a magnesian lime and more careful tests for that should be made. If there is no residue the lime is likely to have a high calcium content. A residue which is sandy suggests a lean lime and a smooth, claylike residue (Redgrave calls this 'greasy or unctuous') suggests a hydraulic lime.

Vicat[5] described a method of classifying limes between fat, lean and various degrees of hydraulicity. This involved preparing walnut-sized lumps of quicklime, dipping them in water, observing the speed or slowness of initial hydration and the heat generated, fully hydrating the samples and then storing them for extensive periods below water in an open-topped jar. The procedure takes a long time but requires very little equipment.

If a lime is described as eminently hydraulic (or any similar description) it must be assumed that builders and designers will expect to place reliance on its strength and for this a conclusive test must be carried out. But for most other properties a degree of latitude is likely to be acceptable.

Conclusion

To get the best out of the wide range of limes available it is necessary to draw up a wide classification of limes.

Only large-scale producers can afford to undertake the types of test required in most Western standards. To assure confidence in limes from small-scale producers it will be necessary to devise helpful field tests which can be used by producers and purchasers. These can be backed up with laboratory testing, but even the laboratory tests should be as simple as possible.

Strongly hydraulic limes used for structural purposes must be proved by testing for the strength of lime-sand briquettes. International co-operation is needed to establish these standards and tests.

27. The Smeaton Project: Factors affecting the properties of lime-based mortars for use in the repair and conservation of historic buildings

IAIN McCAIG*

Interest in the use of lime-based materials – mortars, grouts, plasters, renders and paints – for use in the repair and maintenance of historic buildings and monuments in the UK has been growing steadily during the past 15–20 years. To a large extent, current practices have evolved through trial and error informed by only limited scientific and academic research.

Although there is now a significant body of experience in the UK in the use of lime-based materials, it is apparent that practice is not always matching theory and that there are still many partially unsolved problems.

The characteristics of mortars may be defined in several ways; in practical terms the ones which concern us most are:

a) Fresh mortar:
  - workability
  - rate of hardening
  - shrinkage.
b) Hardened mortar:
  - appearance
  - moisture and air permeability
  - mechanical properties, e.g.
    – adhesion
    – ability to tolerate movements
    – strength
  - durability: resistance to damage by frost and salts.

These, together with the chemical properties of the mortar, must clearly be compatible with existing materials and appropriate to the context in which they are to be used.

The factors affecting the characteristics and behaviour of lime mortars are derived not solely from the mortar constituents but also from the techniques used in processing the ingredients and in preparing and placing mortars, and the conditions to which they are subject during hardening. The properties of materials in contact with the mortar and ambient environmental conditions at the time of placing and hardening of the mortar are also influential.

The objective of the Smeaton Project research programme is to contribute to the understanding of the characteristics and behaviour of lime

---

*English Heritage, London, UK.

mortars by attempting to identify – and where possible quantify – the material and practice parameters which affect their properties. It is intended that this will lead to the production of practical guidelines for specifying, preparing and utilizing conservation mortars in a wide variety of contexts. The Smeaton Project is a joint research programme of English Heritage, the International Centre for the Study of the Preservation and Restoration of Cultural Property (ICCROM) and the Building Research Establishment.

The name of the project refers to John Smeaton, a mathematical instrument maker and engineer who, in 1756, after experimenting with limes and additives for 'water building' (mortar capable of hardening under water), decided to lay the courses of stone for the Eddystone lighthouse in a mixture of pozzolana from Italy and lias lime from the south-west of England. His report on this work was the first of many studies preceding the current EH/ICCROM project which sought to identify and quantify pozzolanic additives for optimum performance.

Background to current project

The Smeaton Project grew out of experimental work begun by English Heritage in 1986 to identify suitable mortars for use in the conservation of Hadrian's Wall in the north of England. Sections of this Roman wall are on high ground and exposed to severe weather conditions. In the recent past relatively strong Portland cement mortars had been used for pointing and repair. This mortar was itself sufficiently durable to withstand the extreme exposure to which the Roman wall is subjected, but it contributed to the deterioration of the wall in a number of ways:

- by concentrating cycles of wetting and drying through the stone faces
- by trapping water, thereby causing leaching of original core mortar and increasing the risk of frost damage, and
- by increasing the likelihood of mechanical damage to stone on removal of high strength mortar in the course of maintenance works.

Recognition of these problems led to an attempt to exclude all cement in mortars used on the Roman wall. Early trials with lime mortars were not completely successful, the mortars having inadequate frost resistance. Whilst their behaviour as a sacrificial material was technically satisfactory, the frequency of replacement required was not economically sustainable.

In some areas of the Roman wall original jointing and core mortars survive; these were sampled and found to contain lime, crushed tile and crushed sandstone, sand and kiln debris. The binder:aggregate ratio ranged from 1:1 to 1:3. Traces of animal fat, probably tallow, were also observed in the samples. Unfortunately, analysis at that time was not carried out to determine the extent and nature of any pozzolanic reactions between the lime and the crushed tile and any other hydraulic components, nor was pore size distribution examined. This is now being done.

Mortar trials at Hadrian's Wall

In 1986, a programme of comparative durability tests on a range of 120 mortar mixes was commenced at Hadrian's Wall. This exercise was designed primarily to observe the comparative performance of the binding materials in the mortars and was not intended at that stage to simulate site practice.

The trial mortars were grouped in three categories according to the type of the principal binder material:

Group A : non-hydraulic limes
Group B : hydraulic limes
Group C : Portland cements

(The sand type and grading was consistent throughout the trial mixes.) The mortar cubes were prepared in Spring 1986 and placed on exposure racks in two locations near the Roman wall. Two further sets of cubes were stored in a ventilated frost-free environment and were subjected to the BRE salt crystallization test in Spring 1987. The samples exposed on site were visually assessed for signs of breakdown after each winter period.

In 1988, selected mortars were used in the consolidation of small trial sections of the wall. The purpose of these trials can be summarized as follows:

- to evaluate the practical issues involved in the preparation of lime mortars both on site and at a remote depot
- to determine the ease, or otherwise, in the use of lime mortars on site and to record any problems encountered and
- to provide realistic and varied test sites for the long term monitoring of performance of the trial mortars.

It is probably fair to say that the Hadrian's Wall trials have posed more questions than they have answered. It is hoped that the subsequent phases of the Smeaton Project, described below, will help to resolve some of these questions.

The Smeaton Project

The Smeaton Project is to be carried out in phases over a 3-year period and involves the collaboration of architects, materials scientists, conservators and craftsmen at English Heritage, ICCROM, and the Building Research Establishment.

The project has been designed to test the conventional wisdom and assumptions about lime mortars by investigating and quantifying the influence of materials and practice parameters on the characteristics and behaviour of mortars based on the non-hydraulic limes which are generally available in the UK. The work, which will involve both laboratory and field research, includes the following areas of investigation:

Materials parameters

Brick dusts and cements
After the requisite literature search and experimental design stage, laboratory testing has begun in the first phase of trials dealing with the effects of the addition of brick dusts or various cements on the performance of lime/sand mortars.

In case of brick dust, an attempt is being made to quantify such factors as optimum particle size, degree of vitrification, and proportion of dust in the mix. Differences in reactivity between freshly ground and stored brick dust will also be explored.

In terms of cement, the study hopes to compare data with the results of tests reported by Ingmar Holmstrom[1] and those at Hadrian's Wall which showed that small additions of cement to lime/sand mortar actually reduced the performance of the mortar and rendered it inferior to pure lime/sand mortar. Because it is common practice in the UK to add very small amounts of cement to lime mortar in conservation work, this is an important area of concern.

The tests being carried out in the first phase of trials are as follows:

a) on the fresh mortar:
   - water content
   - rate of stiffening
b) on the hardened mortar:
   - water vapour transmission
   - compressive strength
   - salt crystallization (freeze/thaw proxy)
   - rate of carbonation
   - XRD to determine extent of lime/silica reactions
   - pore size distribution.

The experimental work on Phase I of the project will be completed in 1992. In addition to these laboratory-based tests, selected samples are to be exposed to natural weathering on a number of sites.

Aggregates
Subsequent phases will address other materials parameters including aggregates. The relevance of aggregate grading, shape and type to mortar

performance is well known and documented in the context of cement mortars but less well understood in lime mortars. The effect of these parameters on lime mortars is to be examined and quantified in terms of workability, shrinkage, carbonation rate, porosity and pore size distribution, strength and durability.

Porous aggregates such as crushed stone, pumice brick, tile and shell appear to affect the carbonation rate and durability of lime/sand mortars. Recent work in Canada[2] has also emphasized the value of brick aggregate in cement/lime/sand mortars as an air entraining medium (as opposed to the use of foaming agents) as a factor in improving frost resistance. These parameters will be studied with the object of exploiting the phenomena in lime/sand mixes.

When pigments are used to modify mortar colour, their effect as fine aggregate on the performance of lime/sand mixes tends to be overlooked. When the percentage of pigment is very small the effects are negligible, but at 5 per cent and over they begin to be significant in modifying mortar performance. Lime/sand mortars moderately loaded with pigment will be tested for shrinkage, carbonation rate, porosity, strength and durability against unpigmented controls.

Additives

A further phase of the project will deal with additives. Water reducing agents, air-entrainers and organic additives (e.g. fats and waxes) have been shown to influence strength and frost resistance. This phase of trials will attempt to answer the questions:

- what are the quantifiable effects of using such additives in lime/sand; cement/lime/sand and brick dust/lime/sand mortars?
- What proportions are appropriate for what conditions?
- What, if any, are the by-product risks?

Practice parameters

Phases of the project dealing with practice parameters will investigate and attempt to quantify the effects of differing techniques in the preparation and utilization of lime mortars:

Preparation of lime putty

Can performance variables be quantified from lime putty prepared in different ways e.g. lime putty from slaking and hydrated lime run to putty by soaking? What are the effects of storage or 'seasoning'?

Blending and preparation of mortar

Quantify performance variables of fresh and hardened mortars in respect of:

- Form in which lime is used:
  - dry mixed hydrated lime/sand
  - lime putty/sand
  - granular quicklime slaked with sand
- Method of blending binder and aggregates:
  - hand mixing
  - mechanical mixing/milling
  - minimum water vs. excess water
- Storage of mortar vs. immediate use:
- Method of 'knocking up':
  - hand processes (optimum duration and technique)
  - mechanical milling (optimum duration and roller/paddle settings)
- Method and timing of introduction of additives.

Utilization of mortar

Quantify performance variables of fresh and hardened mortars in respect of:

- Workability/consistency.
- Compaction of mortar:
  - degree of compaction
  - influence of aggregate type/shape
  - method and timing of compaction (duration and limiting factors)
- Finishing:
  - closed texture vs. open texture.

Setting and hardening of mortar

Quantify performance variables of fresh and hardened mortars in respect of:

- effects of low and high suction backgrounds (to what extent can effects be regulated by pre-treatment of surfaces or by mortar formulation?)
- Environmental conditions:
  - effects of drying rates
  - methods of control
- Other treatments:
  - periodic re-wetting (duration and limiting factors).

The final practice parameters phase will involve the field testing of selected mortar mixes.

The time schedule for the project is as outlined in Appendix B.

Conclusion

The repair of mortar joints, plasters and renders undoubtedly requires an understanding of the materials which have survived and which are to be matched and repaired, together with sufficient knowledge of the materials, which are available now to prepare a sensible specification for such work. We should attempt to find a match as close as possible so that our new material can co-exist with the old in a sympathetic, supportive and, if necessary, sacrificial capacity.

However, there are some important points to be remembered: firstly, our materials are not always the same as those used by the builders whose work we are to repair – even if we call them by the same name.

Secondly, we are not always using materials in the same way. Although analysis can tell us the ingredients of a mortar, it does not tell how the mortar was made or used. Professionals and craftspeople have to rediscover these practices through trial and error. Field experience suggests that the techniques employed in the preparation and utilization of mortars may be of equal significance to their composition in determining their performance. Laboratory research which is not somehow replicated on site is of limited usefulness.

Thirdly, the performance requirements of a repair mortar may be significantly different from those of the original mortar. For example, the walls of a ruined building remain wetter and colder for longer than those of a roofed and occupied building. In that situation the repair mortar might well need to possess greater resistance to damage by frost and slats than the original.

Information obtained from the Smeaton Project will, it is hoped, provide a more informed basis for the selection manufacture and utilization of suitable lime mortars for specific purposes.

Appendix

The Smeaton Project

General outline of programme
            Literature Review
- Phase I   Materials parameters
  - Brick dust and cements.
- Phase II   Materials parameters
  - Aggregates.
- Phase III Material parameters
  - Air entrainers, water reducers and organic additives.
- Phase IV Practice parameters
  - Preparation of binders
  - Blending & preparation of mortars
  - Utilization of mortars
  - Setting and hardening of mortars.

Phase V   Practice parameters
- Field trials on selected mortars
Analysis of historic mortars
- development of techniques for analysis
- characterization of durable and non durable historic mortars
- comparison with replacement mortars.

28. Eurolime: Development and manufacturing of lime for the preservation of monuments

JOHN FIDLER*

This paper is a short report on progress towards co-ordinated European technical research on building limes: lime-based mortars, renders, plasters and shelter coatings, and the manufacturing, processing and construction craft practices associated with them.

'Eurolime' constitutes a family of aligned and overlapping research programmes under the umbrella of the EUREKA EUROCARE research protocols. Countries involved to date include the United Kingdom, Germany, Sweden and the Netherlands with expertise also drawn from ICCROM, the International Centre for the Study of the Preservation and Restoration of Cultural Property in Rome, Italy.

EUREKA is an international research co-ordination and development system founded by the French and German governments outside of, but aligned with, the European Community's operations. It was originally designed to counter advanced technical and scientific research emanating from the United States 'Star Wars' programme and Japan's '5th Generation' computers work. Now, it is designed to stimulate European co-operation and facilitate market-driven research and development on 'new market' products and services, with the minimum of bureaucracy.

Under the EUREKA umbrella, EUROCARE deals with the conservation and maintenance of the built environment. EUROCARE member countries fund their own expenses in running programmes and share burdens equally. Some states *directly sponsor* national research work as a contribution to international programmes, others only help by co-ordinating industrial and academic participation and by steering concepts

---

*English Heritage, London, UK.

to realistic market needs. The United Kingdom is involved in the latter way and English Heritage is responsible for the national secretariat, with Norway providing international EUROCARE co-ordination.

Overall project leadership for Eurolime has been provided through Germany and the Geophysics Department of the University of Karlsruhe. In the United Kingdom, the only significantly funded national research has been provided by English Heritage in partnership with ICCROM, Rome, through its discreet Smeaton Project, described by Iain McCaig.

Work being considered in this preliminary period of Eurolime development involves:

- Mapping national lime types in use in the building industry: advancing on the EC/CEN standards work for air limes and helping to catalogue hydraulic, hydrated and other mixed limes and their chemical constituents and properties.
- Cataloguing common national building practices with lime: to map the discreet differences and similarities in operational work across Europe to see how site storage, the mixing of limes, additives, water and aggregates effects workability, plasticity, initial and intermediate stiffness, flexibility, porosity and rain and frost resistance irrespective of proportion in the mix.
- The effects of pozzolans (trass, pozzolana, pulverized fuel ash, brick dust etc.) on the qualities of lime mortars.

The EUROCARE protocol works as follows:

Interested parties from industry, academic and commercial research houses and other sectors combine at national level and join with colleagues in at least two other European countries to share research ideas and develop a programme. The joint research plan is then developed with constituent members offering resources in money and kind to take the work forward. The project is then certified by the EUREKA system and promoted widely in order to attract other research partners, sponsorship and commercial interest.

The EUREKA protocol in itself does not provide researchers with any national or international funding. However, the EC strategic grants programmes in DG XII for STEP/RTD (Science and Technology for Environmental Protection) and for BRITE EURAM (manufacturing technology) can be bid for and favoured if EUREKA certified. In the EC's top-down grants programme, millions of Ecus of currency are available for defined research.

Eurolime fits nicely into this set of circumstances – being a benign material with greater potential to be used in the conservation process and having wider construction industry benefits if the right material and additives are manufactured and marketed.

For those interested in the Eurolime project, a description is available at the EUROCARE office in Norway and a poster is ready for display at conferences and exhibitions supplied by Karlsruhe University.

29. Lime stabilized soil blocks for third world housing

DR DAVID J T WEBB*

Millions of new affordable homes will be needed by the turn of the century to house the population of the developing world. The majority of these homes will be built using traditional building materials such as reeds, grasses, soils, bamboo, wood and animal skins. If local material such as soil could be improved by making it more durable, or improving its load bearing capacity, it therefore follows that there would be an improvement to the quality of housing which could be of major benefit to the developing world.

During the last 40 years, engineers and architects have been investigating various soil stabilization methods (employing mainly cement) and acceptable building products have been made on improved block-making machines. Other stabilizing materials such as lime, bitumen, cowdung and chemicals have been investigated rather less.

Lime is actually an excellent soil stabilizer provided certain criteria are followed and correct soil selection is made.

This chapter briefly reviews some of the experiences on lime stabilization by the Building Research Establishment (BRE) over the last 25 years.

Principles of soil stabilization for building work

It has been argued that many countries in the Third World have neglected the utilization of locally available building materials, particularly soil, in favour of a wholesale and often inappropriate adoption of western materials and techniques of house building.

Ever since humans congregated in villages, unbaked soil has been one of the principal building materials used. Because of the clay fraction present in the soil, walls built of unstabilized soil swell on taking up water and shrink on drying. This moisture movement gives rise to severe cracking and often leads to difficulties in getting renderings or protective coatings to adhere to the walls, as well as to their eventual disintegration.

*Building Research Establishment, UK.

In the Bible, the families of Noah travelled from the east and found a plain in the land of Shinar and they stayed there. Genesis, Chapter XI, states that: 'and they said unto one another, go to, let us make brick from clay and burn them thoroughly, and they had brick for stone, and slime had they for mortar'. In this context slime refers to bitumen because archaeologists have uncovered some of these early buildings and irrefutable evidence exists that bitumen was used as a building material mixed with soil or sand to form a waterproof mortar.

Several thousand years ago, the Romans stabilized road bases with hydrated lime and these appear to have given outstanding service. The aim of soil stabilization is to increase the soil's durability to the destructive properties of the weather and this may be achieved by one or more of the following methods:

- increasing the strength and cohesive properties of a soil
- reducing the permeability of soil
- making the soil more water-resistant.

Modern manufacturing technology has improved the quality of cementitious materials and chemical compounds to such a degree that the use of a correct stabilizer can improve the compressive strength of a soil by as much as 400–500 per cent. This improvement greatly increases the soil's resistance to erosion, and better standards of housing and living conditions may then be achieved.

Chemical and organic remains which are present in a soil react in differing ways to the types of stabilizer now available.

Therefore the art of soil stabilization is not a precise topic and is based on experience rather than text book formulae.

Methods of soil stabilization

During the last 40 years numerous stabilizing processes have been used, including mechanical stabilization. Chemical additives such as Portland cement, quicklime, hydrated lime, gypsum, alkalis, sodium chloride, calcium chloride, aluminium compounds, silicates, resins, ammonium compounds, polymers, agricultural and industrial waste products[1] have been investigated.

However, because of their cheapness and/or availability in most developing countries, the most widely used stabilizers are Portland cement, lime and bitumen, while gypsum and agricultural waste are used to a lesser extent.

Mechanical stabilization follows on from the adobe method of block making or rammed soil and involves tamping or compacting the soil by using a heavy weight to bring about a reduction in the air void volume of a soil, thus leading to an increase in the soil density and bearing capacity.

Contemporary requirements for roads and embankments have led to the use of vibratory rollers and tampers for soil compaction. Soil compaction is obtained by impact loading and when sufficient compaction has been achieved a pedal is activated which ejects a building block from the mould.

Experiments at the BRE have proved that where a high compacting pressure of about $10N/mm^2$ is applied to either a lime or cement stabilized soil good quality building products are produced. Using this level of compression, considerable savings will be made due to the lower content of stabilizer needed without lowering the resultant of strength of a product.

Other research work has been conducted on the optimum compressive pressures needed on a soil to achieve good durability and to meet acceptable strength standards.

Cement stabilization

Before lime stabilization is discussed, it would be helpful to discuss the cement stabilization of a sandy soil where the sand particles act as a filler. After water is added to the mix, hydration occurs and the soil particles are embedded into a matrix of hard cementitious gel. The small proportion of lime released during the hydration process may react further with the small clay fraction of the soil mix, forming additional cementitious bonds within the soil-cement mix.

For effective stabilization with cement, it is important that the clay fraction is not so high as to swamp the small quantity of cement present. Hence it is inadvisable to employ cement as a stabilizer when there is a high clay fraction present in the soil, and lime should be considered.

Lime stabilization

In the manufacture of hydrated lime, a two-stage method is used. The first stage requires the burning or calcination of limestone (or shells or coral) $CaCO_3$ in a kiln at 900°C. This stage expels carbon dioxide, CO_2, and produces quicklime or calcium oxide, CaO. This product will be in a lump form, white, grey or yellow in colour, which has to be crushed to form a powder.

The second stage involves slaking or hydration whereby a precise volume of water is reacted with the quicklime to produce hydrated lime or calcium hydroxide $Ca(OH)_2$.

Both hydrated and quicklimes can be used to stabilize a high clay fraction soil but, as lime is a caustic material that damages the eyes and skin, careful handling is advised, especially in the case of quicklime which can explode if incorrectly mixed with water.

When lime is used as a stabilizer for soils containing a high clay content, four reactions are believed to occur:

1. Cation exchange. As soon as lime comes into contact with a clay mineral a chemical exchange of ions takes place, giving the clay particles a lower affinity for water. With the subsequent change of the ionic lattice the clay particles will be characterized by a lower moisture movement.
2. Flocculation or agglomeration follows as a direct result of the cation exchange with resulting formation of clusters of microscopically small soil particles, making the mix more viscous or stiff. Flocculation enhances volume stability, increases permeability and decreases plasticity.
3. Carbonation of the lime itself, as it reacts with the carbon dioxide from the air, gives rise to a hardening effect.
4. Pozzolanic reaction is believed to be the most important. It is a chemical reaction between the clay and lime particles, yielding hydrated calcium silicate aluminate compounds similar to some of those found in Portland cement. The rate at which this pozzolanic reaction proceeds is a function of the temperature. Thus it is very slow in temperate climates, but usually fast in the tropics.

A factory installation in Denmark produces building products from a clay-based soil stabilized with lime and moist cured at temperatures between 60° to 97°C for various periods of time up to a maximum of 24 hours.

Cation exchange and flocculation are almost instantaneous as soon as the lime is added to a soil, whilst the last two reactions are much slower, causing the strength of lime stabilized soil blocks to develop over weeks, months or even years.

It has been suggested that when lime is used as a stabilizer instead of Portland cement, the dosage should be doubled. However research at BRE has demonstrated that such doubling is not necessary if a sufficiently high compacting pressure of about $10N/mm^2$ is applied to a high clay content soil. With a higher compaction pressure the following benefits occur:

- volume of air voids is reduced which brings the lime and soil particles into closer contact with one another
- stabilizing reactions occur more readily
- improved density results in durability improvements
- enhanced compressive strengths are obtained.

Factors affecting soil-lime stabilization

Five main factors have been found that affect the way in which soil is stabilized with lime. These are:

1. Soil properties, lime type, lime content, compacting pressure and curing conditions.
2. Some of the soil variables that may affect lime stabilization are particle size distribution, amount and type of clay mineral present, sulphates and organic matter.

3. The clay fraction of a soil sample is defined as that portion which passes a 2 micron sieve.
4. Whilst clays and some silty soils benefit from the addition of lime, sandy soils have been found unsuitable for lime stabilization.
5. Mineralogically and chemically, clay can be divided into five main types, i.e. kaolinite, vermiculite, illite, chlorite and montmorillonite.

Montmorillonite and illite clay minerals consist of a three-layer aluminosilicate compound which has particles weakly bonded together, whilst the other three types have a two-layer structure. The montmorillonite and illite clay minerals normally have high shrinkage rates, up to about 15 per cent, whilst the other clay minerals have lower shrinkage rates of about 8 per cent.

Both types can be stabilized. The high shrinkage rate clay needs an initial addition of sand to modify the mix before stabilizing with lime or cement. The lower shrinkage rate clays need only the addition of lime or cement.

The main influence of clay on soil behaviour is its ability to expand and contract when subjected to changing moisture conditions.

Quicklime appears to be more effective than hydrated lime for enhancing soil strength, possibly because of its greater calcium hydroxide potential. The amount of lime needed for optimum strength under given pressing and curing conditions will vary from soil to soil. For example, Clare & Cruchley[3] report an optimum lime content of 7 per cent to 10 per cent for ten English clay soils when using compacting pressures of $2N/mm^2$.

Experiments performed at BRE have demonstrated that improvements in compressive strength and durability on suitably sized building blocks stabilized with either Portland cement or lime can be achieved when using compacting pressures in the range of 8 to $16N/mm^2$.

Soil selection

In the preparation of a stabilized soil block-making programme, consideration must first be given to how much soil is needed and where a suitable clay content soil can be found.

Small test holes will have to be dug in order to investigate the various depths of the clay and sand layers.

Soils suitable for stabilization should meet the following general requirements:

- The soil should contain sand, clay and silt particles; the sand fraction to form the body, the clay fraction to improve the cohesion qualities and the silt fraction to act as a void filler.
- Ideally the soil should have a good, even, particle size distribution to ensure a uniform homogeneous matrix with a 6mm maximum size.

- After drying, the soil should be readily friable. If not, it may indicate too high a clay fraction and extra sand would probably have to be added to reduce the clay proportion.
- After mixing the soil with water at optimum moisture content, and with a selected stabilizer the soil mix should be easily compacted by hand.

A suitable type of clay soil for lime stabilization is nearly always found just below the organic topsoil strata. The topsoil strata can consist of loamy soil which is used for agricultural purposes. It is often dark in colour, contains roots/plant growth, has a musty smell and is NO GOOD for stabilized soil block production.

Soil testing
There are several tests that can be performed on a soil sample to establish the presence of clay. Three simple site tests can be performed before a more detailed linear shrinkage test is used to determine the optimum lime content.

Hand-moulded cube
After passing a sample through a 6mm sieve, the soil sample is moistened and formed by hand into a cube of about 25mm. If the cube is easily formed there is probably a large amount of clay present in the sample. Allow the cube to dry out in the sun for a day and if surface cracking occurs this will also indicate the presence of clay. If a cube cannot be formed or a cube falls into several pieces after drying out this would probably indicate there is too much sand or silt in the samples.

Rolling test
Remove a soil sample about the size of a chicken's egg and add sufficient water for it to be easily moulded by hand. On a flat, clean surface, attempt to roll out the soil sample into a thread using the palm of the hand. If it is difficult to roll out, more water should be added to the sample.

A considerable amount of information can be obtained from this test if the sample:

i) breaks as soon as it is being rolled out, this indicates too little clay and a high proportion of sand;
ii) is easy to roll down to at least 5mm diameter (the thickness of a pencil), then there is probably enough clay present in the soil for block making;
iii) can be easily rolled to about 3mm diameter (half the thickness of a pencil) then there is a high proportion of clay and little sand in the sample.

Sensory tests
Include the 'Odour test' because organic soils have a musty odour when

freshly dug. Even dry organic soils when wetted will emit a musty odour and it is advisable to discard this type of soil.

The 'Bite test' is a quick and useful way of identifying sand, silt or clay. A small pinch of soil sample lightly ground between the teeth has differing effects. Sandy soils contain sharp and hard particles which grate between the teeth. Silty soils feel smooth to the tongue and clay type soils feel smooth and powdery like flour between the teeth.

The 'Shine test' is performed on a moistened soil by rubbing a fingertip over the surface of the soil. Sandy soil is abrasive to the touch; whereas a clay soil quickly shines and is smooth to the touch.

Linear shrinkage test

This is an important test because the results will indicate the quantity of clay present in the soil and will also be used to determine the amount of stabilizer that should be added to the soil to produce good quality building blocks.

i) Make an open-topped wooden mould with dimensions 40mm × 40mm × 600mm long.
ii) Oil or grease the inside of the mould.
iii) Make a sample of the soil wet enough to form a homogeneous paste which, when tapped, brings water to the surface. This state occurs at the liquid limit of the soil being used. Pack the mould full of this paste and gently knock the mould several times on a hard surface to release any trapped air.

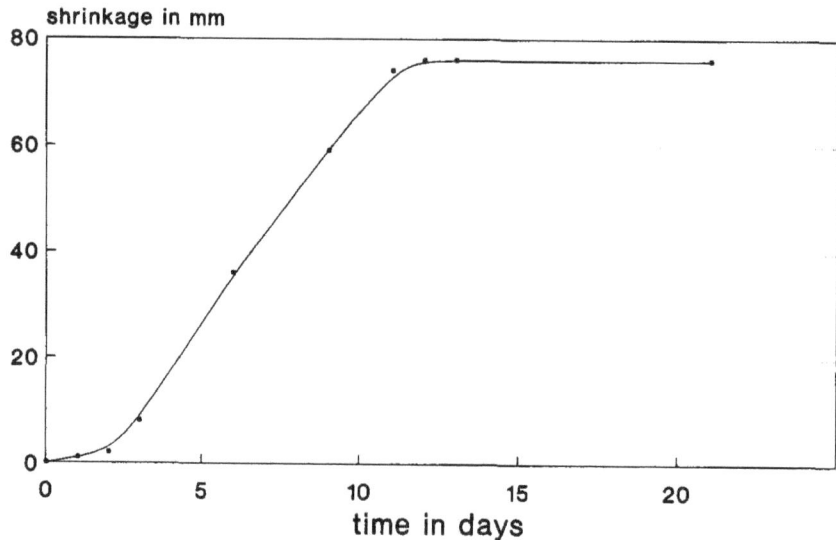

Figure 1: *Typical graph of linear shrinkage/time for high clay content soil from Pakistan*

Table 1. Maximum soil shrinkage

Less than 15mm <2.5%	Not enough clay present. This soil would not be recommended for block-making.
15mm to 30mm 2.5% to 5%	1 part cement to about 20 parts of soil by volume is recommended. Not enough clay present for lime stabilization.
30mm to 45mm 5% to 7.5%	1 part cement to 12 parts of soil OR 1 part lime to 6 parts of soil by volume is recommended.
More than 60mm >10%	Probably insufficient sand, either add sand or it may be worthwhile experimenting with 1 part lime to 4 parts of soil by volume. This method would be expensive when using this quantity of lime and therefore the soil is not recommended for block-making.

iv) Place the mould under cover in a convenient location and regularly take measurements of the shrinkage of the soil sample.

v) The measurement of the shrinkage gap can now be plotted against time as shown in Figure 1 for a typical high clay content soil.

Using a compaction pressure of $10N/mm^2$ the above stabilizer selection table has been based on information in the CINVA-Ram block-making machine field manual and confirmed by BRE experience overseas with only minor amendment.

Block-making machines

CINVA-Ram machine

Many different types of powered or manually operated machines have been developed in the world and the most widely used is the manually operated CINVA-Ram machine which was introduced in 1952 by the Inter-American Housing and Planning Centre in Colombia. The CINVA-Ram block-making machine exerts a low compacting pressure of about $2N/mm^2$ on a stabilized soil mix during the production cycle when making a soil block measuring 290mm × 140mm × 100mm. The soil blocks produced usually have a minimum of 8 per cent cement content (lime cannot be used because of the low compaction pressure) and are cured in the open for about four weeks.

Laboratory trials during the late 1970s at BRE and in Ghana indicated that hydrated lime could also be used as a soil stabilizing agent provided the soil mix is compacted at pressures of between 8 and $16N/mm^2$. The lime content would be about 6 per cent by volume provided the linear shrinkage of the soil is about 7 per cent.

BREPAK machine

The BREPAK block-making machine was developed at BRE in 1981. It employs a 10N/mm² compacting pressure and produces a good quality lime stabilized soil block whose dimensions are similar to a block produced by the CINVA-Ram machine. The BREPAK machine can produce up to 40 blocks per hour, dependent on the number of operatives. Five workers are needed for sieving soil, mixing, machine operation and block stacking. Each block produced on a BREPAK machine weighs about 8kg which is about 20 per cent denser than the CINVA-Ram produced block.

Field trials

The Overseas Development Administration (ODA) funded BRE in the development of the BREPAK machine as well as the first overseas field trial in Kenya of this machine. The object of this field trial was to evaluate this new block-making machine under actual site working conditions. A joint research project was started in 1981 between BRE and the Housing Research and Development Unit (HRDU) of the University of Nairobi.

The joint BRE/HRDU Kenyan collaborative research programme resulted in the construction of some 30 good quality buildings in Kenya using both cement and lime as the stabilizing agent. They achieved cost savings of about 46 per cent when compared with conventional concrete block walling.

The results were achieved when the stabilized soil building blocks were compacted using pressures of 10N/mm² and all the recommendations of the Kenya Standard on stabilized soil blocks were satisfied.

A joint UK/Sudan research project on soil-lime stabilization was first initiated in 1980 by the British Council in Khartoum who sponsored a Sudanese architect from the National Council for Research (NCR) in Khartoum for a study period of two years at BRE.

Three Sudanese soil samples were shipped to BRE and various tests were performed on these soils. For example:

Linear shrinkage 12.2 per cent, 11.3 per cent and 7.9 per cent
Soluble salts 0.58 per cent, 1.46 per cent and 0.74 per cent

Six per cent hydrated lime was used to stabilize these soils and a quantity of test blocks were produced on both a CINVA-Ram and BREPAK block-making machines.

At 27 days, test blocks were immersed in water and, whereas the CINVA-Ram blocks disintegrated, the BREPAK blocks were still sound. They were crushed at 28 days and wet compressive strengths were found to be 4.7, 6.15 and 3.3N/mm². However, long-term durability tests after two years on the BREPAK blocks indicated that good dry strengths were maintained but the corresponding wet compressive strengths dropped by about 15 per cent from the original 28 days wet strengths.

A BREPAK machine was supplied to NCR who sponsored the construction of a 60m² school building with stabilized building blocks using either lime or cement to stabilize the soil.

During August/September 1988, Sudan experienced about three weeks of unprecedented high levels of rainfall centred over the Khartoum and Omdurman areas which caused the River Nile to burst its banks. An extensive area around Khartoum was flooded including the partially completed school building which was standing in about one metre depth of water for about two weeks. After the flood water receded, the school building was inspected for erosion and the blocks used for wall and floor construction were found to be in good condition whereas some 200 000 traditional earthen houses had disintegrated in the flood and some two million people were left homeless.

Following a relief appeal by the Sudan Government, 70 BREPAK machines were supplied to Sudan by the Foster Plan International Organization of America to support the flood relief rehousing programme. Considerable time was taken up by mobilization of this project including many delays in getting plant and materials into the country, and selection and training of operatives.

During the first year of this self-help building project, some 800 houses were constructed from both lime and cement stabilized soil building blocks.

Since 1981, the BREPAK machine has been used in some 25 developing countries including Kenya, Sudan, China, Botswana, Jamaica, Egypt and St Vincent. Numerous requests have been made about increasing the hourly output rate of the machine.

Cartem Elephant block maker

In 1991 the Cartem Elephant block maker machine was introduced onto the UK market. This machine, whilst retaining the BREPAK pressing parameters, has an hourly output of about 130 blocks when manually operated and up to 250 blocks/hour from a powered version.

The Cartem Elephant block maker makes use of a circular turntable with three mould cavities so that mould filling, soil compaction and soil block ejection can all be conducted simultaneously.

Stabilized soil block standards

Numerous standards have been developed covering the use of fired clay products especially in developed countries. However, in the majority of Third World countries, the operation of building product standards is non-existent, especially in the field of soil block construction.

Before acceptable standards can be introduced for stabilized soil building elements, they first must be proven in the field and, as early as the mid-1970s, research in this field was initiated at BRE in conjunction with Ghana which indicated that good quality stabilized soil building blocks could be produced on site.

'CRATerre', the International Centre for Research and Application of Earth Construction in France, in 1984 proposed a draft Sudanese technical standard for lime stabilized compressed soil blocks.

This proposed technical standard was derived from a study of soil block making machines together with field experience gained from working in Sudan.

In 1989, BRE prepared a draft specification for stabilized soil building blocks which was presented to a workshop in May 1989 held at the United Nations Commission for Human Settlements (UNCHS) in Nairobi, Kenya. The participants at this workshop accepted the main proposals of the draft specification and published their findings in May 1990.[4]

The main points from the Kenya Standard Specification for Stabilized Soil Blocks are:

i) The maximum particle size of the soil must be less than 6mm, i.e. Fine Gravel Classification
ii) The soil must be free of deleterious materials such as organic matter and soluble salts, i.e. loss on ignition of less than 12 per cent after a chemical analysis has been performed
iii) The proposed standard be named Portland Cement and Lime Standards specifications
iv) The sum of the alumina, silica and iron oxide elements should be greater than 75 per cent after conducting a chemical analysis
v) The clay and silt content should exceed 10 per cent, as determined from the result of a sieve analysis
vi) Water absorption after 24 hours total immersion should not exceed 15 per cent of the original mass
vii) Specific weight of a block at 28 days to be $1800kg/m^3$ minimum and $2000kg/m^3$ recommended. When freshly moulded, a block to be $1850kg/m^3$ minimum and $2100kg/m^3$ recommended.
viii) The minimum unconfined wet compressive strength after 24 hours water immersion shall be not less than $1.5N/mm^2$. The wet compressive strength shall not be greater than 50 per cent of the dry compressive strength, i.e. F dry => Twice × F wet.

Conclusion

During the last 25 years, the research at BRE on soil stabilization, which has encompassed some 25 developing countries, has proven that good quality lime stabilized soil building blocks can be produced.

The building blocks produced under compacting pressures of about 10N/mm^2 will comply with the recent standards now in operation. Therefore good quality housing can be provided to assist with the housing of those people in the developing world who are homeless.

30. Traditional and current uses of lime mortar, render, and stucco in Zanzibar

FATMA I KARA*

Over many centuries, lime has been used extensively in Zanzibar and the East African Coast in general. As well as in construction (which will be described later), lime has been used since the 18th and 20th centuries in small-scale sugar refineries owned by Arab and Indian merchants, and now in the only sugar factory owned by the government. Lime, due to its corrosiveness, has also been indigenously used to protect stored food from insect attack.

Another agricultural use of lime related to the above is the spreading of it in fields to prevent organisms (like snails and grasshoppers) from attacking plants.

Lime has also been used medicinally in treating cuts and wounds. It used to be mixed with leaves of certain trees (*mbono*) to dress a fresh wound.

Another extensive use of lime is in betel chewing where the advantage of alkalinity is put to use as a stimulant to the mouth of the user. This habit was brought to the islands by the Indians who are still very strongly attached to it.

Production and preparation

Lime in Zanzibar and the East African coast is produced by heating either limestone or sea-shells at high temperatures. It is mainly used in powder form (dry hydrate slaked lime) although in the old times lime used to be buried in pits in the form of putty before being used several years later. Even the mortar itself used to be buried as such before use.

Lime in construction

It is difficult to establish, even approximately, the first use of lime in construction in Zanzibar. Generally lime as a material and its technology is

---

*Stone Town Conservation and Development Authority (STCDA), Zanzibar

attributed to *Wadeburi*, a vague description of a foreign people hitherto unresolved. The latest theory is that they are Malaysian people who travelled via Zanzibar during their pre-medieval voyages to as far as Malagasy. If anything, this shows that lime was used in construction even before the times of building such historical sites as the Kizimkazi Persian Mosque (11th century), Mvuleni and Fukuchani Portuguese residences and forts (medieval) and the Unguja Ukuu Shirazi, old capital of Zanzibar (9th century).

A closer observation of these structures suggests that the present use of lime in construction has come about very gradually.

The earlier Shirazi (Persian) ruins were but small structures like the Kaole tomb ruins in Bagamoyo and the graves scattered all over Zanzibar: Paje, Pujini, Ndagoni, etc.

Even the Kizimkazi mosque which is a relatively small structure could not survive long, to the extent that by the 18th century only the kibla portion (Mihrab) was standing.

The coming of the Portuguese added new dimensions to lime construction technology. Larger, massive structures like forts (Fort Jesus, Old Fort, Mvuleni) were built, until by the time of the conquest by the Omanis, lime construction technology was fully accepted and established. Hence the development of Zanzibar Stone Town and the other East African coastal towns in the form they are seen today.

Applications

The use of lime in construction is very diverse, ranging from foundations to roofing as elaborated below:

Foundations

The structural performance of the many historic buildings proves beyond reasonable doubt that generally lime/stone foundations are safe and sound. Normally the foundations (as all structures) were massive. There seems to be no indication that the lime/laterite ratio for foundations varied from that of walls or slabs. However, there are many situations where the lower portion of such foundations were courses of stone only, a practice now widely adopted in local construction. But lately there has arisen widespread, unconfirmed belief that houses built on lime-mortar/stone foundations are more vulnerable to termite attack than those on cement-based foundations. Perhaps this has also something to do with the less salty condition of the soil compared to the houses within the Stone Town where no such observations exist. The Stone Town was, during its construction, almost an islet.

Walls

The common (imported) use of lime in walls is that of a binder. Normally it is mixed with laterite and sand in varying proportions depending upon many factors: quality of lime, clay-content of the soil, desired background colour, workability, etc. Some well constructed walls are left unplastered and the intended substrate remains exposed revealing its heterogeneous appearance of lime-mortar spotted with various stones randomly placed. In the beach villages where clay is not easily available for the traditional (wattle and daub) mud-wall construction, lime-mortar is used in much the same way as the mud would have been used: as a filler between large stones and pole framework. This exposed framework is often covered both sides with another layer of mortar before finally applying plaster. Presently, a cement-lime mortar is also used.

Suspended floor slabs

The mortar proportions are not necessarily different from that of walls or foundations. Normally the slabs are supported by mangrove poles spanning between walls. However, mortar for slabs (and floors) used to be more thoroughly mixed than that for walls. Special donkey driven mixers were exclusively used. Moreover the laterite had a higher clay content than for other elements. Compaction was also very thorough, and this was traditionally carried out by women who were specially employed to achieve a high compaction and a polished finish. To relieve the lengthy and repetitive process involved in this technique it was usual to follow a certain background rhythm. All this was in order to make the slab watertight. There are, however, a few cases where the slab was self supporting. It is said that in such cases marine-limestone was used instead of coral rock. But considering the many cases where the slabs stand long after the poles have rotted, it is very likely that any stone would suffice subject to a careful and well researched arrangement.

Floor finishes

Though no longer used now, all original floor surfaces in the Stone Town were of lime and sand screeds. No proper testing has been done to ascertain the proportions of the mix, but it can safely be assumed that an ordinary lime-plaster mix would suffice to provide a moderately smooth and hard enough surface to sustain the normal stresses of residential use.

It is, however, clearly noticed that the screeds are mostly pebbles from the seashore and/or sea-shells. Some few buildings in the villages had their floors made from a mixture of lime, clay, sand and cow or donkey dung. The fibrous nature of the dung prevented the formation of surface cobweb cracks. Almost without exception the finished surfaces were smooth, watertight and hard enough. Probably this technique was imported by the Indians as the majority of such houses belonged to them.

Stucco

The plaster was made from wet moulds where lime was used with organic oil-based additives. These additives might have been from many sources judging by the different colours of the stucco in various buildings. Brick dust, which was imported from Oman, and is now also produced locally, was also extensively used, probably even in the Ithnashary Dispensary as suggested by the final colour of the stucco work there. It appears that, in some of the other buildings, gypsum was also added to the lime for stucco work.

Plaster

Lime plaster, usually lime:sand:laterite, is applied in three coats. Originally the skim layer was applied while the undercoat was still wet. The skim layer is of lime putty originally made of finely sieved lime immersed in water for at least a month.

The addition of laterite to the skim layer is likely to cause cracks especially if proper curing is not exercised. The degree of cracks depends, among other things, upon the clay content of the laterite used.

In some modern construction of concrete block walling cement is mixed with lime to provide a background for lime plaster. This is followed by a very thin layer of lime or lime/sand thoroughly trowelled while still wet. It has an added advantage of blending well with paint as many of the available paints are lime based.

Limewash

Almost all the Stone Town houses were originally limewashed. This was achieved by covering lime powder with water for at least a week, and regular stirring. On application, the lime water was poured out before reaching the lime putty below ready for application in two to three coats. Without additives, such limewash is vulnerable to bio-deterioration. Copper sulphate is sometimes added to give it both an off-white colour and act as a fungicide. Sometimes boiled seaweed residue is applied to give it strength against abrasions. Sugar is said to have been added for the same objective.

Current use of lime

Despite being used extensively during earlier centuries, the use of lime in Zanzibar has declined considerably. At present lime is used only for limewash, plasterwork and/or mortar for general use.

This decline has been brought about by the following factors:

i) Use of cement: Many people took the relatively short setting time of cement as a clear advantage over lime. Moreover, the ease and convenience with which one can build with cement and mass produce components (bricks, concrete) convinced many people not to 'think back' to lime. The idea (fashionable trend) that cement as an imported material was much better than lime contributed considerably to the decline in its use.
ii) Mis-treatment: Originally lime was slaked shortly after burning and used for building soon afterwards. There are many lime producers, however, who did not realize that to store dry hydrate in the open for long periods would result in air slaking. This led generally to inadequate performance which resulted in people considering lime a poor binder.
iii) Many master builders and craftsmen fled the country due to political upheavals without adequately passing on their skills to those who remained. Thus trades like stucco-work could no longer be carried out.

This led to the present situation where lime is considered a second choice to cement. In many cases, even the limited application of lime is technically doubtful. In many cases, people use lime just to economize on the use of cement, a combination which only reduces the performance of both materials.

In applying cement to surfaces originally built with lime mortar, the end result is deterioration of the repaired portion as the cement prevents evaporation of trapped water. In Stone Town where rising damp is common and where trapped water incorporates dissolved salts from the marine atmosphere, salt formation below the repaired layers adds impetus to such destruction.

Limewash does not adhere well to surfaces made of cement-based mortars. Now that STCDA is trying to revive the proper use of lime under the advice of ITDG, we anticipate that things will change for the better, particularly due to ever-increasing cement prices.

Conclusion

Proper knowledge in the use of lime was until recently a thing of the past. A great deal needs to be done to return lime technology and use to its original status. We at the Stone Town are just beginning to establish a proper research programme and facilities in these areas. The lack of specific proportions of mixes for particular uses such as stucco, plastering, etc. is due to the non-existence of written, established records from previous generations. Let us hope that with our dedication and co-operation, future generations will not repeat this mistake.

31. Lime: a common binder for preparation of mortar in earth construction

HUGO HOUBEN*

Alternative binders, in particular lime, are appreciated mainly because they require a low investment for their production and can be produced even in small-scale kilns. In earth construction, lime is appreciated for its specific physico-chemical characteristics.

Lime may be qualified as a 'soft' binder, i.e. a binder giving low mechanical performance compared to Portland cement. Low mechanical performance is generally considered inconvenient, because people currently confuse mechanical performance with the durability of building materials. In fact, durability of a building relies to a great extent on the compatibility of different materials used in one construction.

Using a sand-cement mortar with high mechanical characteristics (but with low elasticity) on an earthen wall is a mistake, since an earthen wall is susceptible to dimensional variations caused by temperature and moisture changes. Characteristics of lime (elasticity, permeability to moisture/vapour transmission, low mechanical performance) are much more appropriate for preparing mortar in earth construction.

Before accepting the idea that a binder with low mechanical performance will make a building more durable than a binder with better mechanical performance, one must understand how a mortar and an earthen wall work together.

Definition of a mortar

A mortar is a plastic paste, which gradually hardens, and which is obtained by careful dry, and then wet, mixing an aggregate (e.g. sand) and a cementitious agent or binder(s), (e.g. aerated lime, hydraulic lime, cement), possibly with an additive (a colouring agent, fibres, an inert lightweight aggregate such as vermiculite or perlite, or a chemical composite).

mortar = aggregates + binder + (additive) + water

One specific aspect of earth construction is that soil is in fact a natural mortar. The clay contained in the soil acts as a binder and the inert portion (silt, sand and gravel) as an aggregate.

---

* EAS, the Earth Building Material Advisory Service, CRATerre, International Centre for Earth Construction, Villefontaine, France

Types of mortars

The proportion of the different materials making up a mortar will vary depending on the use of which the mortar is to be put. There are several types of mortar depending on the part of the building where they are to be used and their particular intended function.

Masonry mortar

In horizontal masonry joints, the function of the mortar is to:

- spread the vertical loads exerted in the wall
- compensate for any deficiencies in laying.

In horizontal and vertical joints, it serves to fill the voids between blocks, preventing air and rainwater from passing through.

At the foot of the wall, it has an additional function which is to prevent water rising from the soil by capillary action. Even in this it has to be waterproof and the joint is referred to as a waterproof joint.

In general terms, mortar gives a certain resistance to exocentric or lateral forces by fraction and by bonding the blocks together.

Mortar for plaster

A plaster is obtained by applying one or more thin layers of mortar to the exterior or interior surfaces of a wall or partition; its function is to:

- improve the surface's appearance (smoothness, state of the surface)
- provide durable protection of the wall from weathering (rain, wind erosion, abrasion and impact).

A plaster is typically applied in several coats: one thin coat with a rough surface (coarse stuff) to ensure that the plaster sticks to the wall, a main coat (the body of the plaster) to ensure that the surface is impermeable and to make it smooth, and a thin surface coat (finishing coat) for appearance and to complete impermeability. Each of these coats has a specific function and a particular mortar should be selected to suit each one.

Working surface mortar

This type of coating, which is applied to a flat surface (such as a floor, terrace, pavement) and which forms a wearing layer, aims on the one hand to protect it from repeated abrasion or from blows (traffic, furniture legs, etc.) and on the other to make it waterproof, against rain in the case of terraces or pavements or against domestic water or accidental flooding in the case of floor or ground coatings.

Mortar for waterproof coating

Waterproof coats are a particular type of plaster used to protect surfaces from the effects of flowing water (e.g. for drains) or of regular submersion in water (e.g. inner walls of water tanks). The coat has to be able to withstand the passage of water even under pressure and the abrasion of particles washed down by the water.

The need for mortar in building with earth

In low-cost housing schemes, for reasons of economy, using mortar is quite often called into question both for laying compressed earth blocks and for protecting walls. Before making a decision, it is important to assess whether using mortar is really necessary.

Masonry mortar

The development of semi-industrial presses allowing the manufacture of blocks of very accurate dimensions has led certain manufacturers to introduce building systems which do not require the use of a masonry mortar: so-called dry-laying, either self-adhesive or using an adhesive. The short-term economic advantage is attractive. But laying them remains a manual operation and can therefore never be completely perfect. Using a mortar enables imperfections in laying to be rectified and allows each course to be level. In addition, the bond created by the mortar between the blocks is important to the final strength of the building and tests have shown that the mortar contributes 50 per cent to this. Finally, a mortar will also ensure that the joint is wind-proof and that water cannot infiltrate between blocks. In general, the use of mortar when laying blocks or bricks is therefore recommended.

Protective mortar: plaster

A great many earth buildings, even in exposed areas, are never plastered. It is wrong to think that the durability of a building is systematically linked to the presence of a plaster. The first function of a plaster finish is to improve the appearance of a building. If using a plaster is to be avoided, the building must therefore be attractive in itself. Clearly it will not be attractive if it has been carelessly built from materials of doubtful appearance.

External plaster finishes have an additional function which is to protect the body of the wall from weathering. To avoid the necessity of applying a plaster, either the facade in question must be sheltered from inclement weather, or the wall must be made impervious to weathering. The first of these solutions implies that the problem has been addressed at the architectural design stage, through the use of overhanging roofs, organic or artificial barriers, and suitable orientation. The second presupposes that the building materials or the material making up the wall have been reinforced and made insensitive to the effect of water. At the production stage, stabilization procedures, either throughout or solely for the surface of the walls, are in combination enabling one either to eliminate the need for a plaster or to reduce the stresses it will have to withstand. This simplifies the formulation of the composition of such a plaster and gives a greater guarantee of its viability over time.

Properties of mortars

The main properties of mortars are:

Appearance

The colour of a mortar will depend on the colour of the aggregate's fines and of the binder; the surface texture of a plaster will be obtained by the way the finishing coat is applied: scratched or brushed on etc.

Compressive strength

This depends on the nature and the proportion of the binder and the grain size distribution of the aggregates; hydraulic binders will give higher compressive strengths; grain size distribution of sand is also a parameter for achieving compressive strength (better results are obtained with coarse sand or sand with a well graded distribution).

Adhesion to the surface

This will have a direct bearing on the way in which the plaster behaves over time on the wall. Adhesion increases with the amount of binder used but is also improved by the way in which the surface is prepared, and the conditions and quality of application of the plaster (climatic conditions particularly during drying out).

Compatibility with the surface

The most common instances of incompatibility are linked to differences in moisture permeability and to mechanical performance. When a plaster which is impermeable to moisture vapour transmission is applied to a surface which is permeable, the moisture vapour remains trapped between the plaster and the wall. Beyond a certain point this will condense and inevitably lead to irreversible separation of the plaster. If a mortar with high mechanical performance is applied as a plaster to a surface with lower performance, when stresses occur within the plaster (shrinkage during drying out, for example) there is a risk that this will lead to a deterioration of the wall's surface.

Impermeability

Water, in its liquid state, should not be able to pass through a mortar layer applied as a plaster. Such mortar must therefore be very compact to avoid the presence of capillaries. This compactness is ensured both by an adequate quantity of binder and by the mortar being applied with force. The waterproof nature of the plaster is assured by the greater impermeability of the mortar (amount of binder, compactness), linked to its water-repellent properties and an absolutely essential ability to withstand cracking, even if movement occurs in the support.

Ability to withstand cracking

A crack enables water to penetrate easily; withstanding cracking is therefore vital for the plaster to fulfil its function of impermeability. There are several possible causes for the appearances of cracks:

Movements in the support: Cracking in the support will always lead to cracking of the plaster, especially if this consists of a mortar made with a hydraulic binder. But there can also be very slight movements in the support which are imperceptible to the naked eye, and which are most often due to changes in ambient air temperature or humidity or a change in loads carried by the structure.

External influences: When subjected to sudden changes in heat or humidity, the plaster itself can expand faster or slower than the support; this results in differential expansion.

Internal stress: Plaster is subject to internal tensile stresses when it cures and dries out, which will be linked to the mix proportions used, and external climatic conditions during application and curing. Water should in fact evaporate slowly and regularly.

The ability of a plaster to withstand cracking is linked to its capacity to accept deformation which is measured by its modulus of elasticity, E. The elasticity of a mortar is a measure of its ability to accept deformation without breaking up. The more a material can accept deformation, the lower its modulus of elasticity. The modulus of elasticity of a mortar depends to a great extent on the nature of the binder.

Workability

When it is still in a plastic state during application, a mortar should be malleable, making it easier to use and resulting in a satisfactory appearance. This malleability, or workability, depends on the nature of the binder. Even when Portland cement is used as binder, lime is often added to improve workability of a mortar.

Shock-resistance and surface hardness

The capacity to resist shocks is linked to the hardness of the aggregate and the proportion of binder used which should be high.

The matching of materials to earth walls

Even a cursory understanding of the principles for making up a masonry or plastering mortar, or a protective coat, clearly shows the importance of suiting the composition of the product to the nature and physical and mechanical characteristics of the support. Most often, problems encountered with plasters in earth building are due to the composition of the mortar being unsuitable to the specific nature of the materials, which is either poorly understood or not taken into account.

The characteristics of earth as a building material which need to be taken account are essentially as follows:

Shrinkage during drying and setting

During drying out, the withdrawal of water often results in dimensional shrinkage (0.2 to 1mm/m for compressed earth blocks). It is therefore preferable to wait for a month after completing the masonry work. Waiting longer still also avoids problems due to general movements in the building, i.e. settling which is fairly frequent during the first year of the life of any built structure.

Dimensional variations

Earth is susceptible to significant thermal expansion (0.02 to 0.2 per cent for stabilized compressed earth blocks, compared with 0.02 to 0.05 per cent for concrete blocks and 0 to 0.02 per cent for fired bricks). Swelling due to wetting or any element will vary according to the proportion of stabilizer but can be quite significant. When subjected to a load, an earth wall changes shape perceptibly. Tests have shown that at the height of the modulus of elasticity, an earth wall is four times more flexible (or less rigid) than a cement block wall. The capacity to accept deformation of the mortar must therefore be at least as high as that of the blocks (i.e. it should have a low elasticity modulus).

Mechanical cohesion

Earth has only fairly low tensile strength. Using a mortar with high mechanical performance runs the risk of damage to the earth structure while it is hardening or when differential expansion occurs.

Vapour permeability

Earth is a material which is highly permeable to vapour. If the passage of water in the form of vapour is blocked by a mortar which will not allow gases to pass through it, this will lead to moisture accumulating to the point of saturation, and probably the beginning of deterioration.

Porosity

Blocks will display a tendency to absorb some of the water contained in the mortar. When a mortar has been prepared using a hydraulic binder which requires a certain amount of water for hydration to take place, care must be taken that the wall will not tend to dry out the mortar too early, resulting in its being poorly cured.

Surface state

In principle the sides of compressed earth blocks will be acceptable flat, but with a smooth surface. It is therefore unnecessary, and not recommended, to apply very thick plasters to achieve a final flat surface, but the plaster has to be made to stick to the surface, for example by using a scratch-coat to form a key.

The basis characteristics of earth as a building material imply that in selecting a mortar, preference will be given mainly to those which ensure:

- good ability to accept deformation (i.e. low modulus of elasticity and high ductility, or ability to elongate) to avoid cracking due to dimensional variation
- good vapour permeability
- mechanical performances which are compatible with those of the blocks.

On elasticity and permeability to vapour transit, lime appears as more appropriate than a hydraulic binder (Portland cement or hydraulic lime). As an example, a lime-sand mortar (1:3) shows a vapour permeability of $1.1 m^2.h.mm.Hg$ and a cement-sand mortar (1:4) shows a vapour permeability of $0.50 m^2.h.mm.Hg$, half of the permeability of lime mortar. Regarding the modulus of elasticity, a sand-cement mortar, depending on the composition, is twice to four times more rigid than a sand-lime mortar.

In addition, at the application stage, care should be taken to:

- work on a building only after shrinkage has stopped.
- work only on a surface which has been prepared (i.e. dust removed, moistened, and possibly with a key made by scouring or adding fixing points).
- ensure that there is a bond between the mortar, all the joints are correctly filled in and plastering mortars applied with force.

When deciding on the composition, preparation and application of a mortar for an earth building it is therefore important to bear in mind not only the above recommendations, but also the requirements which will follow on from the way in which the mortar is to be used. Depending on the use of the mortar, these recommendations and requirements are sometimes contradictory and finding a compromise is often problematic.

Masonry mortar

A masonry mortar must have compressive strength and be workable. A thin hydraulic mortar, a sand mortar with aerated lime, or an earth mortar to which coarse sand has been added and which has been stabilized, can all therefore be used. Even a thin earth mortar can be considered for well sheltered walls. It should be noted that the optimum strength of a masonry mortar is in the order of half the compressive strength of built-up blocks.

There is, therefore, no need to use a very rich mortar. Lime, which results in better workability and average mechanical strength, is preferable to hydraulic binders. Only in the case of waterproof joints should one use a very compact mortar based on a hydraulic binder with a high added proportion of water-repellent product wherever possible.

Plastering mortar

The composition of a plastering mortar will vary according to whether the mortar will have to display properties of impermeability or not:

i) Internal plasters only need to have a good adherence and a good ability to withstand cracking, as well as a pleasant appearance. Very simple mortars, such as earth mortars with added sand, weak mortars or aerated lime, gypsum mortars, or lime and coarse gypsum mortars, are therefore all suitable.

ii) External plaster must be compact, in order to be impermeable, and able to withstand cracking, or in other words supple. These qualities are very difficult to find united in a single mortar. This is why it is preferable to apply several layers and to change the composition of the mortar for each layer.

Main body of the mortar

The main layer should be fairly compact to ensure its impermeability and therefore should have a fairly high proportion of binder. At the same time, mechanical performance must remain lower than that of the support. This is why pure hydraulic binders, which would have mechanical strengths which are too high, are not recommended. Mortars made up of a mix of different materials are preferable (e.g. aerated lime and Portland cement) with a high proportion of lime or lime mortars. What is important is to ensure that the modulus of elasticity is below that of the support. The proportion of binder can be lowered, but another solution consists in maintaining the proportion of binder whilst complementing the aggregates with a lightening inert load such as perlite or vermiculite. The thickness of the main layer should be sufficient to prevent the passage of water but it should at the same time not be too thick (i.e. too heavy for the support) especially when it is made with mortars based on aerated lime (in which case, carbon dioxide in the atmosphere can no longer reach through to the middle of the layer and curing is delayed).

Finishing coat

Small cracks are likely to appear after the main body of the plaster has dried out. These can be filled with a final fine and very supple coat. The proportion of binder used in this coat will therefore be lower than that used for the main layer.

Scratch-coat or key

Finally, it is important to ensure that the plaster sticks to the wall; however, increasing the ability of the main layer of mortar to withstand cracking (by amending the proportion of binder), also reduces its ability to stick. An intermediate 'scratch' coat or key is therefore most often used. This does not contribute to impermeability, and cracks which appear can be filled after drying out by the main layer and the finishing coat. The scratch-coat can therefore be made from a mortar containing a high proportion of hydraulic or mixed binder. The only constraints are that the passage of moisture should not be blocked and that too much tensile stress should not be transmitted during drying out. This means that the coat must be thin and discontinuous.

Moisture will be able to escape through cracks and pores which will appear during drying out. On brittle or surface-damaged supports this scratch-coat should be replaced or complemented by the use of fixing points such as broken pottery or nails driven in at different angles over the whole surface.

Working surface mortar

To withstand blows and continuous abrasion, a 'working surface' mortar needs to contain a very high proportion of binder. This high proportion would imply a low tendency to deform. An earth floor, however, is very susceptible to movement (e.g. swelling, or settling). It is therefore highly contradictory to put a rigid working surface on an earth floor. A cement screed on an earth floor will crack and deteriorate very rapidly. It is therefore not recommended to attempt to lay a mortar working surface (e.g. screed) directly onto an earth floor. The solution consists either in making an intermediate rough stone layer between the floor and the screed, or in treating the floor directly by compaction, possibly using a stabilizer, and with regular maintenance, or in ensuring that the surface is protected by another kind of covering (ceramic tiles laid on a bed of sand, for example).

Waterproof mortar

This type of mortar must have an even higher proportion of binder than the preceding type in order to be compact and to ensure that it is totally impermeable. In addition, the use of a strong hydraulic binder would make it impermeable to the passage of moisture. Its rigidity would not allow it to resist inevitable movements in the earth-built structure and it would crack very quickly allowing water to pass into the structure. This kind of mortar has therefore to be both very supple and impermeable which is a very difficult combination to achieve. The fact that it is impossible to reliably protect a drain or cistern made of earth explains why building them in earth is not recommended unless the question is thoroughly understood and unless certain fairly specialized products, natural or artificial, are available.

32. The Landmark Trust: the use of lime-based mortars and plasters in historic building projects

JOHN BUCKNALL*

The Landmark Trust was founded 26 years ago by Sir John and Lady Smith to rescue historic buildings of quality that were either not grand enough to be preserved by such bodies as the National Trust, or too awkward in their location or design to allow for straightforward repair and reuse. Almost 200 buildings have been rescued over that period, and about 150 are now let as holiday accommodation which provides them with a long-term future, and a regular income.

The repair and adjustment of these buildings to new use is in the hands of architects experienced in historic building repair and in traditional building techniques. Nevertheless it must be stressed that modern building technology is also used where appropriate and modifications do have to be made to the buildings to provide for their future.

The buildings vary widely in type from small cottages and follies to manor houses and 19th century forts. A few buildings are of international importance such as the Villa Saraceno by Palladio in the north of Italy, or associated with famous people such as the home of Rudyard Kipling in Vermont. Just as they vary widely in their architecture and historic

---

*Landmark Trust, UK

character they are also extraordinarily diverse in the methods of their construction. The Landmark Trust owns buildings built of cob, stone, timber frame, and rendered brickwork, and they are covered with a wide variety of roofing materials, from heather thatch on the Scottish border, to clay pantiles in Italy.

It is true to say that only in the past 15 years or so has the architectural profession, and the building industry in the UK, begun to recognize the central importance of lime and its supremacy over substitutes, particularly Portland cement. Indeed those very modern materials that claim to be substitutes are all ineffective because they do not have the wide ranging versatility of lime which, used in conjunction with appropriate sands and other materials, provides the essential matrix for mortars, plasters and decorative finishes on countless buildings.

I shall illustrate the use of lime in Landmark projects by providing a number of case studies, but it must be stressed that this is but a small cross-section of the Trust's work.

Villa Saraceno, Fitchanser, North Italy

This very beautiful building was designed by one of the greatest architects of all time, Andrea Palladio, for the Saraceno family in 1545. He illustrated it in his classic work on architecture, *I Quattro Libri dell' Architettura* (Venice 1570).

Construction is of brick covered with plaster both outside and inside, known as 'intonaco'. It is vernacular in character but the architecture is of the highest order. The building is on low-lying fenland and therefore highly susceptible to the effects of damp. The plaster of the villa is almost entirely original, i.e. of the 16th century. The climate in the Veneto is severe but the plasters are of a high quality and, though weathered and damaged, have survived remarkably well.

The plasters are made from slaked semi-hydraulic lime and from sands dug from the site itself. We had proposed to use sand from the site in the repair work but tests have shown it to incorporate too many salts.

The repair of the early plasters is a matter of great interest and importance. In the UK it would be unusual to remove lichens, but the lichens in question at the villa exude acids which are harmful to the lime. The plasters are therefore being washed lightly to remove excess organic growth following which exposed brickwork will be infilled with new areas of intonaco to the original thickness, or simply treated with a light limewash coating to integrate sensitively with the original surfaces. The former treatment will apply where the surrounding intonaco is still relatively thick, and the latter where it has worn very thin. It is to be noted that no additional water is added other than that contained with the lime putty itself as the experience in Italy is that this will cause severe shrinkage on drying.

The internal plasters are also of great interest. The encrustations of centuries of over-painting are being removed to reveal a very beautiful white finish to the original plaster. The friezes in the two great chambers were richly decorated with frescoes, but were covered in dirt and salts. It was considered that large areas of the frescoes had vanished, but the application of trial poultices and the cleaning of two large sections of frieze in the sala have proved that, generally, the frescoes are in very good condition.

The project has not developed yet to the point where a full analysis and report can be given, but it is true to say that lime technology in Italy is a fine art, and much can be learned about its virtues, and the methods of application from practitioners in that country.

Gurney Street Manor, near Cannington, Somerset

This is an important 15th century manor house which declined in function from the 16th century onwards and, when acquired by the Landmark Trust, had been divided into numerous residential apartments. Fine 15th century roofs, a complete chapel, a number of good fireplaces, and many other original details had survived.

Construction is of rough local sand and limestones which, because of their relatively poor quality, were always covered with a protective render and limewash. Only quoins and copings, together with chimneys, were executed in finer quality limestone. In the building's interior, all walls were plastered, and a small and beautiful late 16th century plaster ceiling survives on one landing.

At the beginning of the repair project, lime pits were dug and lime was slaked into a series of pits for mortars, plasters and limewashes, etc. A mortar mill was constructed to facilitate the very thorough mixing of mortars and plasters, which is essential to the use of lime. On any major project, the setting-up of a full system for the use of lime such as that at Gurney Street is absolutely essential. It should also be said that this arrangement has provided lime for use in plasters, mortars and finishes on a number of other Landmark projects in the west country.

Lime has, of course, been used in all mortars at Cannington, but of great interest is its use in plasters, both in the form of rendering over stonework and on various forms of sub-strata on partitions old and new. One example is the four-coat treatment of plaster for use on masonry walls. Because the masonry is rough, several dubbing out coats are required. This is a relatively coarse plaster incorporating hair, which has to be allowed to dry fully before the second and third coats are applied. If insufficient time is given for drying, severe shrinkage will take place in subsequent coats, leading to crazing. On relatively damp sites like this, drying will take a considerable period, and the need for this must be stressed unless agents are used to accelerate the process. The finishing coat is of a fine putty sand mix, wood floated.

When plastering on to internal partitions it was traditional to use either wattle and daub or cleft laths finished with hair-lime plaster. A variety of techniques can be used, some of which have been used at Cannington, but all of which have been experimented with on this site. These are as follows:

- *Cleft laths* which are made of either oak or chestnut. The roughness of the laths provides a good key for the plaster but it is essential that the spacing is carefully calculated to be appropriate to the coarseness of the plaster concerned. If care is taken, an excellent key can be obtained.
- *Sawn laths*: these can also be of either oak or chestnut but, as the surface is often rather smoother and the edges are uniform, it is absolutely critical to get the spacing between the laths correct. If they are too closely spaced, the key will not be adequate and the plaster will fall away. This is particularly disastrous on ceilings.
- *Reed lathing*: both reed and straw have been used as a traditional key for plaster. It is a particularly familiar technique in the low-lying land of Somerset where reed was readily obtainable. In the English Midlands, it is commonly used in conjunction with the laying of lime ash floors. Considerable pressure has to be applied to make sure that the lime presses through to form an adequate key. Alternatively, the render coat must be sufficiently fine to pass through the straw or reed, to ensure a satisfactory key.
- *Stainless steel mesh*: this is a popular modern substitute for lathing and is extremely effective but it must be stressed that the stainless steel is very expensive. It must not be assumed that it is the least expensive option that can be adopted. In many countries, natural wood on straw lathing will be much cheaper.
- *Stainless steel lath*: this is far more effective than the plain mesh as the lathing is designed to permit an excellent key.

Generally speaking, the traditional laths have been used in the repair of all ancient partitions at Gurney Street.

Woodsford Castle, Dorset

This is a fragment of a substantial medieval castle with a vast thatched roof to the exterior, the largest in England. All areas of surviving medieval plaster have been retained and new lime plaster of comparable specification and texture has been used on the other surfaces to give a harmonious whole.

Cawood Castle, Cawood, Yorkshire

This castle preserves the character of the interiors where fragments of medieval plaster have been retained and the whole has been limewashed. It is also important to note that limewash is taken over the stonework of the windows in the medieval manner.

Timber-frame buildings

Sir John Smith has likened the repair of timber-frame buildings to the patching of cobwebs, such is their complexity and the need to retain as much as possible, not only of the frame structure, but of surviving ancient infill panels.

Ty Mawr

This is an extraordinary ancient Welsh house built in 1400 which the Landmark Trust is still hoping to be able to repair in collaboration with CADW, the Welsh Historic Building Agency. An area retains fragments of its medieval infill panelling of wattle and daub, and a very primitive plaster – probably a mix of cowdung and straw.

St Winfred's Well

This is a remarkable survival of a late 15th century timber-framed chapel built over a sacred well chamber fed by a spring. After the Reformation it became a court house and then in the late 18th/early 19th century it became a cottage in which use it has remained until the present day. The decision was taken to repair part of the building using the traditional technique of staves, bent laths, and lime hair plaster.

Internal surfaces of the building have either been lime plastered or limewashed, but the oak has been left undecorated as is the tradition in this part of the English countryside.

East Banqueting House, Chipping Campden

This building was constructed as a banqueting house in the 1620s. The house was extremely dilapidated when acquired by the Landmark Trust.

Lime mortars were used for the repair of the high quality limestone stonework. With this material, it is possible to repair fine quality detail without loss.

Lime plasters were used throughout the interior. Mr Jeff Orton ran the fine moulding of the upper chamber ceiling. Limewashes have been used to all surfaces.

Swarkestone Pavilion, Derbyshire

This small pavilion, from which gentlemen of the 17th century and their ladies looked out upon sports in the garden below, was roofless and dilapidated when acquired by Landmark from the HarpurCrewe family in the early 1980s.

Despite its roofless state, it was remarkable in retaining much of its internal plaster. On analysis this turned out not to be a conventional lime plaster but to have been made of crushed alabaster, and formed into a plaster of extraordinary hardness and durability. Alabaster is a form of pure gypsum, and a French gypsum plaster was used to repair it, as it proved impossible to reproduce the original specification.

This project is included as it illustrates the need always to be cautious in assessing the nature of early plasters before assuming that they are of a particular well known type.

South Street Torrington

This is a fine house built in 1701 in the small town of Great Torrington in north Devon. This district was renowned, from the 16th to 18th centuries, for the high standard of decorative plasterwork carried out in many buildings. This fine quality lime plasterwork was largely executed *in situ*. The front parlour of the house, which is shortly to be repaired by Landmark, has the most extraordinary musical theme with musical instruments to full life-size. It demonstrates the sophistication that plasterwork had reached in England by the early 18th century.

The Pavilion, Ingestre

Built in the mid 18th century, there remains but a fragment of a larger building. Landmark have repaired it and built an extension to form a new living room and bedrooms.

The stonework has been repaired using lime plasters, but the most remarkable feature is the plaster decoration to the vault and back wall of the portico. In the end, about 75 per cent of the original plasterwork has been left intact when the most straightforward technique would have been to replace it entirely. Once repaired, the plaster was given a very thin limewash finish.

The Bath House, Walton

Near Wellesbourne in Warwickshire, the Bath House was designed by Sanderson Miller in about 1750, and was decorated with shell work. The walls are decorated with festoons of shell work and the dome is decorated with stalactites or icicles.

The walls were replastered using lime hair plaster on stainless steel lath set free of the stonework. The shell work was restored by matching shells obtained from the West Indies, with impressions left in the original plasterwork.

This unusual exercise was carried out with a team of specialists brought from a number of different firms, thus combining their individual expertise into a highly rewarding programme of work.

Conclusion

The Landmark Trust has derived great pleasure from working with a remarkable team of architects in the UK and abroad who have, in turn, introduced us to excellent practitioners in the appropriate trades of bricklaying, stone masonry, and above all, plastering. Lime is at the centre of the science and art of building, otherwise Vitruvius would not have given it the prominence he did in his own writings. The processes discussed by him, Palladio, and architects and builders through the centuries are as relevant to us today in all parts of the world as they were then.

33. Use of lime: some techno-social considerations

N G DAVE* and S K MALHOTRA*

Although lime has more diverse uses than any other material and is probably one of the most highly utilized naturally occurring substances, its first image all over the world is that of a building material. This is in view of the fact that lime has had a rich heritage in construction. It is an excellent building material and has proved its worth through age-old structures that have triumphantly braved the onslaught of centuries. Lime is endowed with several superior properties that are not to be found in other materials. Consequently it has numerous uses, and it has been stated that 'without lime, life on this planet would have simply ceased'.

Manufacture of lime in India[1]

Presently the burning of limestone in India is carried out in kilns of various shapes, sizes and designs (which range from the ultra-modern to the very hackneyed types); and operates from the scale of a sophisticated chemical processing plant down to cottage industry scale. As might be expected, under this situation, different types of units exist, but the majority of them consist of very small cottage industry units, which produce a large proportion of the nation's output.

---

*Central Building Research Institute, Roorkee, India

A vast majority of the Indian lime kilns can be classified as vertical shaft mixed-feed kilns and operate on a counter-current principle. Three different designs are quite popular:

- the rectangular country type
- the stemless funnel-shaped conical type
- the cylindrical type

Cottage sector

The rectangular country type of kilns are quite popular in certain areas. They are usually rectangular box-type structures, with a big opening (which is plugged during operation) on one side and a number of air-holes on two opposite sides. These air-holes face each other and are also connected through flues made on the floor of the kiln. The walls of the kilns are thick and are constructed out of locally available materials. These kilns are usually operated in batches, one run lasting as much as three weeks or more.

The stemless funnel-shaped conical type of kilns are popular in certain parts of the country. They, as the designation suggests, are in the shape of a stemless funnel. They are also constructed of locally available materials. The thickness of the kiln walls changes from top to bottom. In certain interior parts they are constructed out of stone and the inner lining is made with 'green' bricks made out of the locally available clayey soil; these are baked *in situ*. These kilns are normally operated on a semi-continuous basis.

The two kiln types above constitute what is known as the cottage-scale sector of the Indian lime industry. They consume different fuels, many of them local, depending on availability. These include cinder and low-grade steam coals. Some of these kilns have been investigated by the Central Building Research Institute and have been found very inefficient, both thermally and in production. The material produced by these kilns is also of inferior grade in most cases. The efficiencies of these kilns are of low order and consequently the utilization of fuel and burning methods are not good, resulting in the generation of polluting gases.

Small-scale sector

In other parts of the country, cylindrical types of kilns are more popular. These kilns are also of various shapes and sizes, but are usually strengthened with several buttresses. They are normally constructed of locally available stones; firebricks are only rarely used. Some of these kilns have ingenious devices for the supply of air as well as for the discharge of lime. They are also run on a semi-continuous basis.

The units produce 10–25 tonnes of lime per day. This is the sector which is responsible for most of the lime produced in India and is concentrated in about 25 important places in various parts of the country.

These lime manufacturing centres have the advantage of being within the vicinity of limestone deposits as well as other favourable economic factors. The kilns used by these units are operated in a more efficient manner.

Lime kilns at four of these centres have been investigated in detail by the Central Building Research Institute. The results have indicated that these kilns also leave much to be desired, and their designs as well as operational practice need considerable improvement. The industry is still traditional in character and almost no advantage has been taken of recent technological advances. The fuels used are mostly steam coals of various types and contain large quantities of volatile matter as well as ashes.

Captive lime production units constitute the third part of the lime industry in India. Captive production is mostly for the chemical industry and, in accordance with the higher requirements of purity and quality of material demanded, sophisticated designs of kilns are employed for production. The fuels used are also of a better grade and in several cases liquid or gasifier fuels are utilized. Consequently there is better utilization of the energy supply.

The Central Building Research Institute, Roorkee has been engaged in research and development of lime kilns for more than two decades. Following a detailed survey of typical kilns at important lime manufacturing centres, systematic investigations have been carried out at the CBRI on the masonry type of vertical mixed-feed lime kilns. As a result of this work the technology for designing and setting up the masonry mixed-feed lime shaft kilns of improved design, for capacities of production up to 15 tonnes of lime per day, has been developed. Many of the kilns have been working on a commercial scale for over two decades.

Pozzolana

Many of the limes are highly calcareous and do not develop good hydraulic strength. The addition of pozzolana is required in such cases. There are two main types of pozzolanic materials, naturally occurring and artificially produced. The naturally occurring pozzolanas are generally produced due to certain phenomena taking place inside the earth's crust. India is poorly supplied with naturally occurring pozzolanic materials and has to depend upon those which are artificially produced. Amongst these are burnt clays, fly ashes, rice husk ash and railway wastes.

In India, burnt clay pozzolana can be produced by calcining suitable soils. The three different technologies used for their manufacture are:

Down-draught kiln[3]

The principle of down-draught has been utilized for developing a new design to produce burnt clay pozzolana. This technique has been found to be quite suitable for small-scale operation. The clay is manually moulded in the form of bricks; these are sun-dried and loaded into the kiln. The kiln is fired in the normal manner and the bricks are allowed to soak for a period of 3–5 hours at the optimum temperature. The temperature inside the kiln is controlled, with the help of thermocouples. Each batch takes 1–2 days for completion of firing and about 4 days for the entire cycle. The product can then be ground to the required fineness.

The process is suitable for use in rural areas where unskilled human labour is available at inexpensive rates. Even a 5-tonnes per day plant has been found to be economical.

Fluidized bed kiln[4]

The process of fluidization, wherein a bed of granular material is agitated by a current of gas passing through it to exhibit some properties of fluids, has successfully been applied for the firing of clays to produce pozzolanas.

The clay sample is sun-dried, disintegrated and fed into the kiln. As the material falls due to gravity, it comes in intimate contact with the up-draught of the hot flue gases generated by the oil burners. The material passes through the burning zone (where a predetermined temperature is maintained), allowed to cool and then pulverized.

This process is continuous and the product is of uniform quality. The capital investment is low. In spite of the use of a costly liquid fuel, it is possible to produce highly reactive clay pozzolana at a fairly low cost. The process is especially suited for non-plastic types of soils.

Vertical shaft kiln[5]

Another system for the manufacture of burnt-clay pozzolana makes use of the vertical shaft kiln. These types of kilns are highly efficient thermally and are largely used for lime manufacture in India and other countries. The technology has been extended for the manufacture of clay pozzolana too.

A rather tall, brick-lined vertical shaft kiln is used for this purpose. Clay in the form of dried clods of 5–10cm size, mixed with slack coal, in required quantities, is fed into the kiln. Air is introduced with the help of a blower and the required temperature is maintained for the stipulated duration. The kiln operates on a semi-continuous basis and the cooled, discharged material is ground to the required fineness.

This process is highly economical and thermally efficient and material quality can be well-maintained. It, however, suffers from the limitation that only highly plastic soils can be calcined in this manner.

Surkhi[2]

One very commonly available pozzolanic material in various parts of the country is surkhi, which is produced by grinding broken bricks and other wastes of the brick kilns, such as over-fired and under-fired materials. Because of the variable nature of these materials, the properties of each batch vary, and this results in the development of poor strengths, when these surkhis are used in mortar or concrete.

Fly ash

The use of pulverized fuel ash, or fly ash, as a pozzolana for use in mortars and concretes has become well established in many countries. India is one of those countries where these ashes are produced in huge quantities. In as much as the chemical composition and the properties of the coals burnt vary, as well as the efficiencies of the process of combustion, not only are the samples obtained from various plants subject to variation but some range is probable in samples collected from the same plant over a period of time.

The Indian fly ashes[6] in general are very finely divided and most of the samples are actually finer than Portland cement. Some of the samples also contain a fair share of unburnt carbon. However, the major constituent, as elsewhere too, is glass, with quartz, mullite and iron oxide as the chief crystalline components.

Rice husk ash

Paddy is one of the major crops of India. In the rice production process, husk is also produced. About 20 million tonnes of the material has been estimated to be produced in the country.

The husk of rice, besides the organic matter, consists mostly of silica. If the conditions during the firing of the husk are correct, this silica can acquire enormous surface area and thus possess very high lime reactivity values. Consequently it is a pozzolana of very good quality. Ensuring proper conditions is essential, otherwise, if any sintering takes place, the surfaces can shrink very fast, and the quality of the pozzolana is reduced.

For producing pozzolana from rice husk, several processes have been developed. In one of these the rice husk is converted into balls with the help of some clay and these balls are then fired in a type of clamp.[7] The result of grinding yields pozzolana, which is claimed to be of a very high order.

In the second process rice husk is incinerated under controlled conditions without admixtures.[8] The ashes obtained have been claimed to be highly reactive and produce mortars with very good properties.

In the third process, a mixture of rice husk and lime sludge is converted into the shape of balls or bricks and then incinerated in a clamp.[9] The resultant material is an integrated mixture of lime and pozzolana and performs quite satisfactorily. This process has the additional advantage of producing lime and pozzolana utilizing two waste materials.

Use of lime in construction

Lime can be used in mortars, plasters, foundations and other situations. In lime based binders, lime imparts all the good properties like workability, plasticity, water retention, autogenous healing and bonding which are essential for masonry work.

Lime can be used with other pozzolanic materials and OPC in various proportions. The various mixtures used are lime:sand, lime:pozzolana, lime:cement:sand, activated lime:pozzolana mixture and rice husk:ash mixtures. A number of mixtures of varying proportion have been recommended in standards and Codes of Practice for the standardization, testing and application of these mixtures.[10]

In recent years research work has been directed towards development of rapid setting and quick-hardening lime:pozzolana binders. Under quick-setting lime:pozzolana binders, the activated lime:pozzolana mixture[11] (ALPM) and rice husk:lime mixtures[12] (RHLM) with improved setting and hardening properties have been developed at this Institute and these binders have been manufactured and used satisfactorily.

Air pollution

As observed above, the lime industry is distributed widely throughout the country but there are certain manufacturing centres where there tends to be a concentration. About 25 such manufacturing centres exist and each of them produces several thousand tonnes of lime every day. As happens in all industries, small as well as bigger centres were established near towns and cities. With the growth in population cities and towns have grown and engulfed these manufacturing centres. On the other hand, various types of kilns are not working in the best possible manner, because manufacturing techniques, as well as the quality of the raw material are not ideal. This results in the release of pollutant gases and particles into the atmosphere.[13] The suspended particles and gaseous hydrocarbons constitute the bulk of released products. A large part of discharge gases are carbon dioxide as well as some sulphur dioxide and carbon monoxide. The magnitude of these constituents varies from place to place and kiln to kiln, yet all of them are hazardous to the health of the local population, animals and plants.[14]

Environmentalists are worried about the effects of these gaseous releases on health and growth in society. There is clearly a requirement for pollution control systems to be established for the kilns. However, the question remains whether the small/cottage-scale industries can bear the burden of this additional cost.

The Central Building Research Institute, Roorkee in collaboration with the National Environmental Engineering Institute, Nagpur, has measured the effluent of the kilns in several places and are engaged in developing low-cost devices for the control of the pollutants.[14] Technologies developed in other countries for abatement may be useful for these kilns but cost is the problem.

Lime stabilization

Soil stabilization is a well known technique for improving the strength and durability of soils, and has also been put into use for the construction of houses and roads. In 1948, four thousand soil cement stabilized houses were built in Punjab (India) for displaced persons from Pakistan; this was considered to be the biggest project of its kind in the world.

Recent studies at this Institute and elsewhere have proved the usefulness of lime and lime:fly ash as a successful soil stabilizer.[15] Bricks produced with lime or lime:fly ash have sufficient strength for use in the construction of one or two storey buildings. In this area of road construction, investigations have been carried out on the use of lime:fly ash mixtures. Lime:fly ash concrete can be used as a sub-base or base course in pavements and as precast blocks for footpaths. Lime:fly ash, sand (or moorum) bound macadam may be used as a base course for lime:fly ash soil as a sub-base course in pavements.[16]

Cellular concrete

Cellular concrete[17] is a lightweight material and can be produced by autoclaving a mix of fine siliceous material (such as ground sand or fly ash) and a binder like cement or lime. The use of this material leads to greater productivity in building, reduction in transportation cost and lowering of foundation loads. The construction cost of these blocks is considerably lower than conventional cement-based cellular concrete blocks. The material is ideally suited for the construction of multi-storeyed structures.

Sand-lime bricks

Calcium silicate bricks, popularly known as sand-lime bricks, are made from sand or a siliceous waste and hydrated lime. In view of the heavy requirement for traditional clay bricks in India there has been a need to develop an alternative building element like calcium silicate bricks. Manufacture of these bricks is economically feasible where sand or siliceous waste and lime are readily available, and good quality clay is not found.

Based on the studies carried out at CBRI[18], good quality calcium silicate bricks of compressive strength of 100–200kg/cm² and water absorption of 10–20 per cent have been produced by using different raw materials.

Energy advantages

Among the various building materials, cementitious materials are of considerable importance. A very large percentage of these binders, up to as high as 45 per cent, are consumed as mortars and plasters. The heat consumption for production of various building materials is given in Table 1.[19]

In masonry construction generally cement, lime, and pozzolanic materials in different proportions have been used with sand and water.

The energy consumption of various types of mortars and plasters have

Table 1. Energy consumption of various building materials

S. No.		Energy consumed (Kcal/kg)		
		Solid fuel	Electrical	Total
1.	Lime	722	0	722
2.	Hydrated lime	547	50	597
3.	Fly ash	0	0	0
4.	Surkhi	0	50	50
5.	Burnt clay pozzolana	228	56	284
6.	Rice husk ash	0	50	50
7.	Activated lime-pozzolana mixture	181	50	231
8.	Rice husk ash:lime mixture	175	50	225
9.	Portland cement	1600	103	1703

Table 2. Energy consumption of various mortars and plasters

S. No.	Type of mortar	Proportion (by volume)	Energy consumption (1000Kcal/m³)	Compressive strength 28 days (kg/cm²)
1	Cement:Sand	1:3	720	50
2.	Cement:Sand	1:6	395	30
3.	Lime:Cement:Sand	1:1:6	448	30–50
4.	"	1:2:9	345	20–30
5.	"	1:3:12	243	7–15
6.	Lime:Surkhi	1:2	261	15–25
7.	Lime:Fly ash	1:2	200	15–20
8.	Lime:Burnt clay: Pozzolana sand	1:1:6	193	30
9.	"	1:2:4	286	58
10.	ALPM:Sand	1:2	142	35
11.	"	1:3	107	25
12.	RHAM:Sand	1:3	64	85

been computed and are reported in Table 2.[20] It is obvious from these computations that mortars containing cement are high in energy consumption. The addition of lime to make composite mortars decreases the energy input considerably. Lime:pozzolana mixtures consume a much lower quantity of energy. The recently developed lime-based materials like ALPM and RHAM are highly energy efficient.

Employment generation

One of the targets of any welfare society is to provide equitable employment for the whole population. In every society there are people with different types of background and experience. To sustain themselves they require jobs commensurate with their background and experience.

All industries require limestone (or lime) at one stage or another as their principal raw material. Fortunately it occurs widely and is distributed throughout the length and breadth of India. There is hardly anywhere that limestone, in some form or another, is not available. Important industries have been listed in Table 3.

The utilization of labour starts from collection and production of the limestone itself. Limestone occurs as mineral deposits in a large number of

Table 3. Important industries using lime and limestone

Limestone	Lime
Cement	Metallurgical processes
Iron and steel	Chemicals manufacturer
Non-ferrous metals	Sanitation
Sugar refining	Pulp and paper
Paper manufacture	Food and by-products
Fertilizer	Dairying
Foundry, glass, etc.	Leather processing
Calcium carbide	Petroleum refining
Soda ash	Rubber
Caustic soda	Protection coatings
Bleaching powder	Paints and pigments
Flux	Bricklaying
Acid neutralization	Masonry mortar
Aggregate (in construction)	Plaster
	Acoustical
Filler	Acoustical whitewash
Agriculture	Soil stabilization
Other uses	Soil liming
	Lakes and ponds

places and quarrying is carried out for the procurement of this material. After blasting the loosened material is collected and transported. These operations require manual labour in the shape of blasters, loaders and drivers.

One alternative source of limestone is Kankar.[21] In most parts of the country deposits of nodules and calcareous material occur with the incorporation of impurities of argillaceous/siliceous material. This kankar is exposed during agricultural operations and is extracted and transported. Often the deposits do not contain over-burden and are naturally exposed. In these cases digging can start from the surface.

Limestone is also found in the form of sea-shells in coastal areas. At numerous places large deposits are available and can be exploited. Often the collection operation is under water and sound transport organization is necessary to collect and carry them for further processing.

The manufacture of lime from these three types of deposits – quarried material, kankar and sea-shells – requires different types of processing equipment, whereas quarried material can be burnt in kilns of numerous different types. Kankar and sea-shells can be fired in very small kilns of special design. Limestone-burning kilns used in India are mostly countercurrent vertical shaft types. These usually have a production capacity in the range of 5–20 tonnes of lime per day. At many places, smaller kilns are also being utilized. Most operations for this type of kiln are by hand and consequently require a considerable amount of manual labour. Some mechanization is being introduced in the larger kilns; however, this is limited. These kilns belong to the cottage-scale sector.

The lime industry is highly labour intensive. Manual labour is required right from quarrying to finished production stage. Labour is also required even after the lime has been produced including transport for its use on site. It has been estimated that about 70 000 people are employed directly in the manufacturing operation.[22]

Despite the present numbers employed to meet production targets, lime production will have to be increased further to meet the shortage of cementitious material. It has been estimated that the investment required for production of lime/lime-pozzolana mixtures is in the order of Rs200/tonne as compared to an investment of Rs1000/tonne required for the production of cement. It is obvious that the investment for the setting up of lime/lime–pozzolana industries is much lower. All investments are highly employment generative.[23] The situation in most developing countries of Asia, Africa and America is similar to that discussed above. In many places where limestone and local fuels are available, small/cottage-scale industries can be established.

Social problems

Besides causing unemployment and pollution, the development of a centralized lime industry has created other social problems. Earlier, the small-scale intermittent type of kiln operation required attention only during charging followed by several days of rest and then discharging. However, present industry requires continuous working and labour has to be engaged in a series of different shifts.

Most labour working lime kilns used to be female. Obviously, in the new scenario this main source of labour can operate only during daylight hours. Consequently male labour is becoming scarce and costly. The outcome is that there are tensions followed by possible strikes and other labour discontent. The result of this situation is that it is now proposed to introduce mechanization on a larger scale in the lime industry. Greater mechanization means a further reduction in the use of labour (which causes more unemployment). At the same time, there is a greater need for electricity, steel and materials requiring capital intensive investment. Rapid and increasing mechanization therefore places undue stress on scarce resources where small-scale, less centralized development would be possible without making such demands.

Housing the millions

The increasing cost of building materials and the increase in population has put a very great stress on housing for the people. More and more people in cities, towns and villages need appropriate roofs over their heads, while prices are sky-rocketing all the time.

For any type of construction a binding material is required and lime is one such material which can find use in small villages, towns and larger conurbations. As the cost of production and transport is low, lime and its products can be available locally and in nearby areas. It can also be processed with the help of economic local fuels. As a result, a low-cost binder can be produced, even it if is sometimes below the generally accepted standard. Lime-based products can also be used for making blocks of various shapes and sizes. These can be produced at local level and thus a nucleus for low-cost housing construction becomes available.

It is for the local communities in various areas to become involved with these aspects of small scale lime production and use to generate housing for the millions.

Appendix 1

Introduction

The following Appendix gives details of lime-based mixes used in actual building work.

Appendix 1a describes the mixes used in conservation work on a centuries-old building known as the East Banqueting House which is situated in the English town of Chipping Campden. Two recipes and methods of plastering are presented – one applied to metal lath, and the other applied directly onto stone. Metal lath is normally attached to material to be plastered where the material is too weak to take a plaster coat, where the surface is too uneven, or where a poor bond between that plaster and the material is expected.

Appendix 1b gives recipes for plaster mixes traditionally used in Italy. The author has re-created these types of plasters and has made samples of each of the types of plaster described.

By using the highest quality materials in mixes like those described and the services of skilled artisans highly intricate decorative stucco work or extremely smooth and attractively coloured plaster finishes can be achieved. The plasters and stucco themselves could last for many centuries without the need for any major repair work. This shows the great versatility of lime, which has at times (mistakenly) been labelled a 'poor man's cement'. In addition to being suitable for use in the construction of relatively simple buildings, lime can be used in the most complex and intricate decorative work on buildings of national importance.

Appendix 1a: Lime plastering specification
JOHN BUCKNALL*

East Banqueting House, Chipping Campden, carried out by Trumper Bros. Ltd (1989–90)

Plain plastering onto metal lath

Backing coat
Coarse stuff: 3 parts sharp washed sand to 1 part lime putty (haired for the pricking up coat).

The coarse stuff was gauged, immediately before use, with class B plaster. On the barrel vault ceiling the proportion was 5 parts coarse stuff to 1 part plaster. For the other ceilings and stud partition walls the proportion was 3 parts coarse stuff to 1 part plaster. The plaster was added to reduce the waiting time between coats.

Finishing coat
Setting stuff: 1 part fine sand to 1 part lime putty (putty was pushed through a 20 mesh sieve).

This was applied in three coats, with the first coat applied in two successive layers. The first layer was gauged with class A plaster retarded with glue size and applied as tight as possible. Straight away the second layer was skimmed over the first. The gypsum plaster in the first layer improves the adhesion and gives a more even surface to work off by controlling the suction.

After a brief waiting period (about 30 minutes) the second coat was applied in the opposite direction to the first (a steel trowel or a wood/sponge float can be used for this).

After another brief period the third coat was applied, again as tight as possible, in the same direction as the first.

The whole of the work was then scoured with a sponge float and trowelled, then brushed twice in opposite directions with a broad flat brush.

---

*Landmark Trust, UK

Plain plastering onto stone walls

Backing coat
Coarse stuff: 2½ parts sharp washed sand to 1 part lime putty, well haired.

Applied as tight as possible following the contours of the stone, cross-scratched with diagonal lines using a wooden lath scratcher to form a key, and left to stand as long as possible (at least a week).

Finishing coat
Setting stuff: 1 part fine sand to 1 part lime putty (putty was pushed through a 20 mesh sieve).

This was applied in the same way as for the metal lath work but without the gypsum plaster in the first coat. The gypsum was not included because of the risk of attack from hydroscopic salts.

Materials

Metal lathing:	Stainless steel 'rib-lath' and expanded metal lath (supplied by 'Expanded Metal Co. Ltd') fixed using stainless steel nails and stainless steel tie wire.
Sand:	Sharp washed sand (Tilcon Ltd)
	Bagged concreting sand (Merridon Pit)
	Fine sand (British Industrial Sand, Grade 50)
Lime:	Slaked lime putty (H.J. Chard & Son Ltd)
Gypsum plasters:	'Hemihydrate, class A' i.e. Plaster of Paris (British Gypsum Ltd)
	Grade – Fine Casting Plaster
	'Retarded hemihydrate, class B' (British Gypsum Ltd)
	Type – Thistle Board Finish
Hair:	Goat hair (Trumper Bros. Also available from H.J. Chard & Sons and others)

Appendix 1b: Traditional Italian lime stucco recipes
J R ORTON C.R.P.*

Base for samples

Sample number 1 is formed on wood lath.
Samples numbers 2 to 7 are built up on clay tiles (to represent the wall).

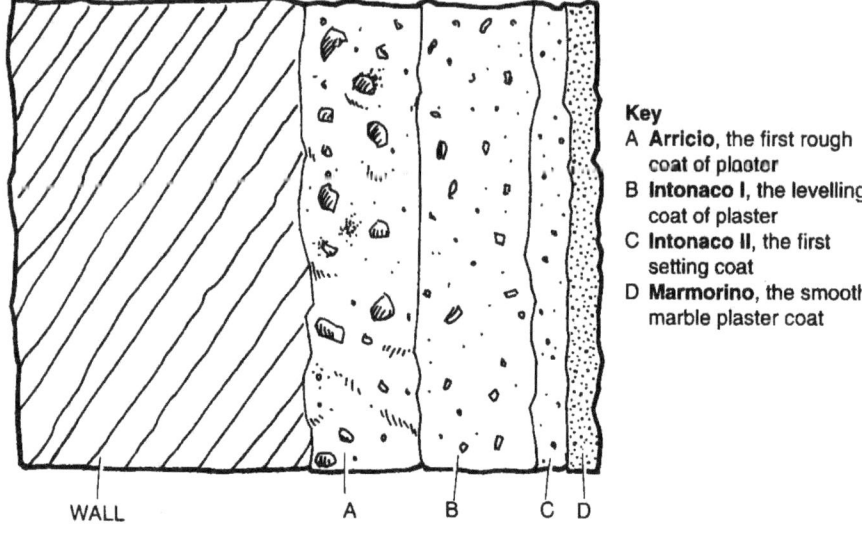

Key
A **Arricio**, the first rough coat of plaster
B **Intonaco I**, the levelling coat of plaster
C **Intonaco II**, the first setting coat
D **Marmorino**, the smooth marble plaster coat

Figure 1: *Section showing the build-up of coats for sample 6, Plain Marmorino (the other samples are built up in a similar manner)*

Mixes for the samples

Note: numbers in brackets refer to suppliers and numbers followed by M.S. are the mesh sieve size; i.e. 20 M.S. refers to 20 mesh per inch.

1. **Ceilings, frescoed**

    a) **Arricio**
    1 part coarse crushed soft brick (4) 8 M.S.
    1 part slaked lime putty (1) 20 M.S.

*Master Plasterer and Stuccoist, Melton Mowbray, Leicestershire, UK
(With additional research on recipes from Rome by Neil Harvey)

b) **Intonaco I**
      3 parts moderately crushed soft brick (4) 12 M.S.
      1 part slaked lime putty (1) 20 M.S.

   c) **Intonaco II**
      1 part fine washed sand (7)
      1 part fine slaked lime putty (1) 40 M.S.

   d) **Fresco**
      Natural earth pigments mixed with limewater applied on the same day as the third coat (10)

2. **Coloured Marmorino**

   a) **Arricio**
      3 parts coarse crushed terracotta (6) 8 M.S.
      1 part slaked lime putty (1) 20 M.S.

   b) **Intonaco I**
      1 part moderately crushed terracotta (6) 12 M.S.
      ½ part sharp washed sand (2B)
      1 part slaked lime putty (1) 20 M.S.

   c) **Intonaco II**
      1 part fine washed sand (7)
      1 part slaked lime putty (1) 40 M.S.

   d) **Marmorino**
      1 part fine marble dust (8) 50–60 M.S.
      1 part fine slaked lime putty (1) 40 M.S.
      Desired colour, (natural earth pigments) (10)

   e) **Decoration**
      as d) – one or more coats

3. **Plain Intonaco**

   a) **Arricio**
      3 parts coarse sharp washed sand (2A)
      ½ part crushed brick (3)
      1 part slaked lime putty (1) 20 M.S.

   b) **Intonaco I**
      3 parts sharp washed sand (2A) 12 M.S.
      1 part slaked lime putty (1) 20 M.S.

   c) **Intonaco II**
      2 parts fine washed sand (2B) 20 M.S.
      1 part slaked lime putty (1) 40 M.S.

d) **Limewash**
   Lime putty (1) 100 M.S. with skimmed milk and pigments (10)

4. Pastellone

a) **Arricio**
   2½ parts coarse sharp washed sand (2A)
   ½ part crushed brick (3)
   1 part slaked lime putty (1) 20 M.S.

b) **Intonaco I**
   3 parts sharp washed sand (2A) 12 M.S.
   1 part slaked lime putty (1) 20 M.S.

c) **Intonaco II**
   2 parts moderate to fine crushed terracotta (5) 20 M.S.
   1 part slaked lime putty (1) 40 M.S.

5. A pozzolan stucco (artificial)

a) **Arricio**
   2 parts coarse sharp washed sand (2A)
   1 part crushed brick (3)
   1 part slaked lime putty (1) 20 M.S.

b) **Intonaco I**
   3 parts pozzolana (coarse to fine crushed brick) (3)
   1 part slaked lime putty (1) 20 M.S.

c) **Intonaco II**
   2 parts pozzolana (moderate to fine crushed brick) (3) 30 M.S.
   1 part slaked lime putty (1) 40 M.S.

6. Plain Marmorino

a) **Arricio**
   3 parts coarse sharp washed sand (2A)
   ½ part crushed brick (3)
   1 part slaked lime putty (1) 20 M.S.

b) **Intonaco I**
   3 parts sharp washed sand (2A)
   1 part slaked lime putty (1) 20 M.S.

c) **Intonaco II**
   2 parts fine washed sand (2B) 20 M.S.
   1 part slaked lime putty (1) 40 M.S.

d) **Marmorino**
   2 parts fine marble dust (8)
   1 part fine slaked lime putty (1) 40 M.S.

7. Ancient Roman stucco

Trial specification for seven coats (total thickness 1¼ inches)

a) **'Trullisatio'** (first rough coat)
   3 parts coarse sharp washed sand (2A)
   1 part slaked lime putty (1) 20 M.S.
   Goat hair.

b) **'Arenatum'** (3 sand coats)
   i) 2 parts coarse sharp washed sand (2A)
      1 part crushed brick (3)
      1 part slaked lime putty (1) 20 M.S.
   ii) 2 parts fine washed sand (2B)
      1 part crushed brick (3)
      1 part slaked lime putty (1) 20 M.S.
   iii) 1 part fine washed sand (2B) 20 M.S.
      1 part fine crushed brick (3) 20 M.S.
      1 part slaked lime putty (1) 20 M.S.

c) **'Marmoratum'** (3 marble coats)
   i) 2 parts coarse marble dust (9) 30 M.S.
      1 part slaked lime putty (1) 40 M.S.
   ii) 2 parts medium marble dust (9) 40 M.S.
      1 part slaked lime putty (1) 40 M.S.
   iii) 2 parts super fine marble dust (9)
      1 part fine slaked lime putty (1) 40 M.S.

Appendix 2
The role of the cementitious binders advisory service (CAS) in disseminating information on lime and alternative cements

OTTO RUSKULIS*

When the Building Advisory Service and Information Network (BASIN) was set up, it was recognized that cements and binders play an important role in all aspects of building construction, and that their production contributes significantly to the economic development of countries and regions. By virtue of its previous project experience in the area of cements and binders in developing countries, ITDG was therefore invited to operate the Cementitious Binders Advisory Service as its contribution to BASIN.

CAS was established in 1988, and has been fully involved with the activities of BASIN since then. One of its first tasks was to begin a search for existing information on cements and binders. This information is being compiled on a computer database which is co-ordinated with the other specialist advisory services on building materials production and use operated by other members of BASIN. Both the search for information and its compilation on computer database are continuing.

Together with establishing a database and information centre on cements and binders, CAS also operates a technical enquiry service. The numbers of enquiries answered increased from about 100 in 1988 to about 200 in 1991. Although these enquiries cover all aspects of building materials and construction an increasing number, now about 30 to 40 per cent are specifically on cements and binders, probably reflecting an increased awareness of the existence of CAS. Further increases in the number of enquiries handled in coming years are expected. Increasingly, CAS is becoming involved with publication of books and technical briefs about cements and binders.

As far as is known, no other specialist information and advisory service on cements and binders exists anywhere in the world, as far as developing countries are concerned. CAS held an international seminar in December 1991 as part of a move to increase its effectiveness and to provide an opportunity for experts to share up-to-date information and experience not only with each other but also, through this book, with a much wider group. It is important that CAS is aware of developments in the field of cements

---

*CAS, ITDG, Rugby, UK

and binders, and in this respect the development of networks of producers, users and researchers in the field is important in identifying problems and solutions, and reducing unnecessary duplication of research and development work. Coinciding with the development of networks, CAS will need to become a more effective dissemination vehicle so that organizations and individuals interested in cements and binders can find out the latest information relevant to their enquiries and, if necessary, be able to contact an organization with some expertise in their own country or region.

Technologies of interest to CAS

In short, CAS is interested in the production and use of all types of cements and binders which have existing and potential application in developing countries, but concentrates particularly on lime and so-called alternative cements. An order of priority of importance with which CAS regards particular cements would be as follows:

1. Lime, whether of the high calcium, hydraulic or dolomitic type
2. Lime-pozzolanas
3. Portland pozzolana cements
4. Portland cement and its production in small-scale plants
5. Gypsum
6. Sulphur and other miscellaneous inorganic binders
7. Asphalt and bitumen
8. Gums, resins and other miscellaneous binders.

Of the above categories, 6 to 8 are considered to have only specialist applications in the building and construction fields, and gypsum is only important as a construction material in certain developing countries, mostly those with extensive resources of rock gypsum and having a relatively dry climate where gypsum can be used externally. Nevertheless, increased utilization of gypsum should be promoted, where possible, especially in view of its relatively low burning temperature in production compared with lime and Portland cement, an important factor in the context of energy conservation.

CAS is particularly keen to promote lime and pozzolanic cements above Portland cement, wherever this is technically and economically feasible. Over the past 50 years or so, Portland cement has become almost the universal cementing material, and traditional cementitious materials such as lime and lime-pozzolanas have tended to be neglected. Traditional skills in their production and use have, to a certain extent, been lost.

Nevertheless Portland cement, in a developing country context, does have a number of disadvantages. Firstly, because of the high capital cost of the plant for producing Portland cement, its production in small units is at a relative cost disadvantage compared with production in large plants. In fact, in only very few areas would production of Portland cement at less than 20 tonnes per day be viable. Much of the world's cement is now produced in large plants and largely for urban markets. In rural areas, where wages are lower and where transport of Portland cement adds further to its cost, this product becomes prohibitively expensive and almost beyond the means of, for example, the small self-help builder. Furthermore, some countries lack the foreign exchange to update or expand their Portland cement plants, and there is an increasing gulf between supply and demand. This tends to increase the cost of cement further and lead to scarcity of supply, particularly in rural areas. Foreign exchange is also needed for spare parts and for experts to assist with major repair and maintenance. This implies that even with the cement plants which are working, few are working at or near their full capacity.

Lime and lime-pozzolana cements can be used in many applications where Portland cement is used and in some, for example in rendering and mortars, they can offer additional advantages. For example, lime-based floors have a disinfecting effect, lime-based renders allow walls to 'breathe', and allow interstitial moisture to be removed before it can damage the building fabric. When lime is used in mortars it improves the ease with which they can be worked. Lime-pozzolana cements and hydraulic lime can even be used in mass concrete. The only applications to which these materials would be unsuited are reinforced concrete, and high or medium strength concretes.

Lime and lime-pozzolana production units, unlike for Portland cement, can be small scale, even of less than one tonne per day, and can be viable in rural or even remotely populated areas. They can normally be built using local skills, with many materials also available locally. Unlike Portland cement, production of lime, or lime-pozzolana cement, is labour rather than capital intensive, so its production contributes to local economies and encourages local development. Because production plants can be started relatively easily, they can be responsive to the needs of the area, so shortages should not occur. Also, lime production plants can be run seasonally, allowing workers and operators to work on other activities, such as agriculture. In rural areas, the cost of locally-produced lime or lime-pozzolana is likely to be a fraction of the cost of Portland cement.

In addition to compiling information on lime and pozzolanic materials, CAS also compiles information on production of Portland cement by mini-cement plants (20 to 50 tonne per day production), as well as on the uses of Portland cement, and Portland cement-based products. However, it does not give the latter such high priority as the former for the reasons outlined above. It should also be recognized that mini Portland cement plants are not the universal answer to overcoming cement shortages, but are ideally suited to relatively small raw material deposits, which would be unable to support a medium or large-scale plant, and/or sited in self-contained areas with a population of at least 50 000.

Analysis of the proportions of enquiries on cementitious materials received by CAS during the first nine months of 1991 are of interest in indicating which types of cements there is most interest in:

Lime 54 per cent,
Pozzolana 32 per cent,
Portland cement 12 per cent,
Other (gypsum and bitumen) 2 per cent.

This clearly indicates the strong interest in lime and alternative cements. Unfortunately, this interest does not extend to government departments, banks, donor agencies, and many established architects and building contractors. Given more resources, CAS would also wish to influence such persons and organizations in favour of alternative cements, where possible.

Direction of CAS and its place within BASIN

The reasons for Portland cement largely replacing lime and lime-pozzolana cements include: large-scale promotion of Portland cement by cement companies, and an unjustified belief that Portland cement is superior in all cases to alternative cements. In addition, an attitude has developed amongst architects and building contractors to specify Portland cement without considering alternatives. Building standards and specifications on using Portland cement have proliferated, while those on lime and lime-pozzolana cements have largely fallen out of use.

Coinciding with decreasing interest in lime-based cements, some traditional skills in making up and using such materials have been lost. Examples of such lost skills include production and use of hydraulic lime, making good quality lime putty, lime-based concretes, and certain waterproofing and strengthening additives to lime.

In recent years, two types of activity have renewed interest in lime. These are rural development projects, mostly run on a small scale and by NGOs, and building conservation using traditional materials and techniques. The former is probably motivated by the high cost and erratic supply of Portland cement in these areas. Much of this activity, however, has been going on in a fairly unco-ordinated way with little co-operation between producers and users and a lack of published material on project results. Comprehensive reports which have been produced are often not publicized and therefore little known.

CAS sees an important role for itself in compiling information on developments in cement and binder technology and, when appropriate, encouraging the spread of information on such technologies. This should lend credibility to the 'alternative cements' and also reduce the risk of repeating costly research and development work in projects. However, this cannot be done without the co-operation of experts and workers in the field of cements and, particularly, a willingness to produce technical reports and provide them for general distribution. The formation of networks of producers and users is seen as being particularly significant in boosting the wider acceptance of alternative cements.

CAS is itself part of a wider network, BASIN, whose members specialize in building material technologies of particular relevance to developing countries. Through BASIN, CAS can reach a wider audience than by itself. In particular, since ITDG predominantly has projects in only six countries, its knowledge of developments taking place in other countries is limited but, as a whole, BASIN has quite extensive worldwide coverage. One important way in which BASIN disseminates information on appropriate building materials technologies is through its newsletter, *Basin News*. This is currently available free of charge and its circulation continues to increase.

The importance of networking to CAS

CAS is part of the Mineral Industries and Shelter Sector which undertakes diverse activities of which cements and binders are only part. Cements and binders are a broad field and an organization such as ITDG can only hope to cover a proportion of this field in direct project work. For example, ITDG's main project activities in this field are currently in lime, with very little on pozzolanas and nothing on gypsum, Portland cement produced in mini plants, or any other types of binder. To obtain information on these other cements and binders, CAS needs to rely on past reports and, often, reports of other organizations.

More valuable are contacts with direct and relevant experience, for example existing members of ITDG's Building Materials Panel. It is, of course, impossible to question a report, so the information you are given is all you can obtain whereas it is possible to question an expert and gain a broader and deeper understanding of the subject.

SKAT, which operates the Roofing Advisory Service (formerly the Fibre Concrete Roofing Advisory Service) has also recognized that, even with a subject such as fibre-concrete roofing and micro-concrete roofing, a much more specific technology than cements and binders, it is impossible for one organization to have experience in all aspects of it. They have actively taken steps to establish networks of producers and users as well as researchers and promoters, and encourage the participation of project partners and collaborators.

If a technology is to develop and bring benefits, two sets of circumstances must be avoided:

- Control of the technology by only a few organizations since they can use it for their own benefit, exclude outsiders, and prevent the technology giving economic and social benefits, slowing down the process of development.
- A wide diversity of groups and organizations working with the technology with little or no co-ordination between them, since research and development work is often duplicated using up valuable resources unnecessarily. It is often difficult for organizations to see potential for the technology beyond their own project activites. Again development of the technology is stifled.

Arguably, with lime and alternative cements at present, the situation is tending more towards the latter.

The happy medium is, of course, when there are a large number of users of the technology who co-operate closely or are, at least, aware of the work of others and of recent developments. However, in practice, this will not happen unless active steps are taken to encourage this; for example, by one or more organizations taking the lead and attempting to discover what is taking place worldwide, by the formation of networks, and by a willingness and means to publish news of significant developments and innovations. Regular meetings and seminars are a particularly important means of informing about recent developments, identifying common problems and offering solutions to these problems.

CAS sees the establishment of networks as a particularly important way of encouraging the development and spread of lime and alternative cement, especially between developing countries. As well as its links with the BASIN network there are two other networking activities with which it is involved:

- The ITDG Building Materials Panel and
- ITDG Country Offices.

The former is a group of technical experts and advisers covering a large part of the building materials field. The panel operates by holding regular meetings and individual members are available to give advice to CAS and ITDG as a whole if required. In addition a number of members have written publications for ITDG and/or assisted with project work. Although the panel has experts on a wide range of building materials, for example clay bricks, bamboo and timber, a significant number of members have expertise and interests in the cements and binders field. ITDG country offices are now taking an increasing role in the development of the organization in a number of countries, and in Zimbabwe, Peru and Kenya they are actively involved in the building materials field. They offer opportunities to CAS in that they have good local knowledge of developments in cements and binders within their own country. Recently, the three country offices mentioned have begun to set up their own databases based on the BASIN model, which will be co-ordinated by CAS, and this should help them reply to technical enquiries from within their own countries or regions. CAS is especially keen to encourage technical co-operation on cements and binders between ITDG country offices.

CAS also wants to encourage the establishment of other networks, both formal and informal. Existing ITDG country offices could act as the focus of networks within their own countries and regions, the aim being that these would act as models for organizations in areas where ITDG is not represented in establishing similar networks. The aim is to include both producers and users as well as practitioners in the building conservation field within each network to give a broad and balanced representation within a particular network. In particular the knowledge obtained on lime and alternative cements from the building conservation field would also greatly benefit small-scale producers and users in developing countries.

CAS also aims to increase its role in dissemination through publications and, if resources allow, undertaking small research projects of common interest. Both of these activities will be of greater benefit if it can draw on the experiences of other practitioners. Particularly important will be the identification of potential areas of research where information is currently lacking or where current practice is unsatisfactory and could be improved.

Appendix 3
Creating a network: The model and experiences of FAS at SKAT

SUSANNE PREISWERK*

At the beginning of the 1980s, fibre-concrete roofing technology (FCR) was introduced in various development projects. Successes and failures reported to SKAT made us very curious about the technology and we decided to find out whether to promote this technology or not. In 1985/6 about 200 working days financed by the Swiss Development Corporation (SDC) were invested in assessment of the potential of FCR technology. As the results were very positive, a concept for proper technology transfer and its worldwide dissemination was worked out. To achieve this aim, it was decided to build up a network to guarantee information flow, to co-ordinate FCR activities and to create synergies. The concept for a network

---

*Rural Engineer, SKAT, Switzerland

in four phases was presented to SDC. The three-year project for building up this network was then financed by SDC.

The objectives of the FAS network are to provide:

- producers with know-how for the establishment of new production plants
- producers with the latest know-how and findings of the FCR panel, mainly concerning technical rationale, management and marketing
- producers with troubleshooting and technical assistance if problems or mistakes occur during production and application
- the FCR panel with monitoring information and with new facts and experience arising from research and development.

First phase
Several organizations, e.g. GATE, ITDG, ATI, DEH, GRET, select or establish a regional centre of their choice. Jointly with this regional centre, they establish new pilot projects or collaborate with existing FCR producers. The role of such regional centres would be to:

- provide FCR producers with planning help for the establishment of new production plants
- provide producers with the most valid findings of the FCR panel, mainly concerning technical rationale, management and marketing
- help the producers with continued technical support and with troubleshooting if problems or mistakes occur.

The organizations co-ordinate their planning and monitoring of the regional centres and pilot projects within a FCR panel to be set up. This panel consists of the representatives of the donor organizations and of the FCR project group (or part of it). The panel designates the persons who are responsible for technical contact with the regional centres and producers and for their provision with latest technical know-how, technical and management assistance as well as troubleshooting. The producers become associate members of the regional centres according to guidelines which have to be established by the FCR panel.

Second phase
The regional centres take over the role of disseminating FCR know-how and assisting producers. The FCR panel has completed its initial task. Now it takes a background role as an information pool and an enquiry and answer service centre.

Third phase
The regional centres communicate with each other. One of the regional centres takes over the role as information and enquiry answering service from the FCR panel.

PHASE 1

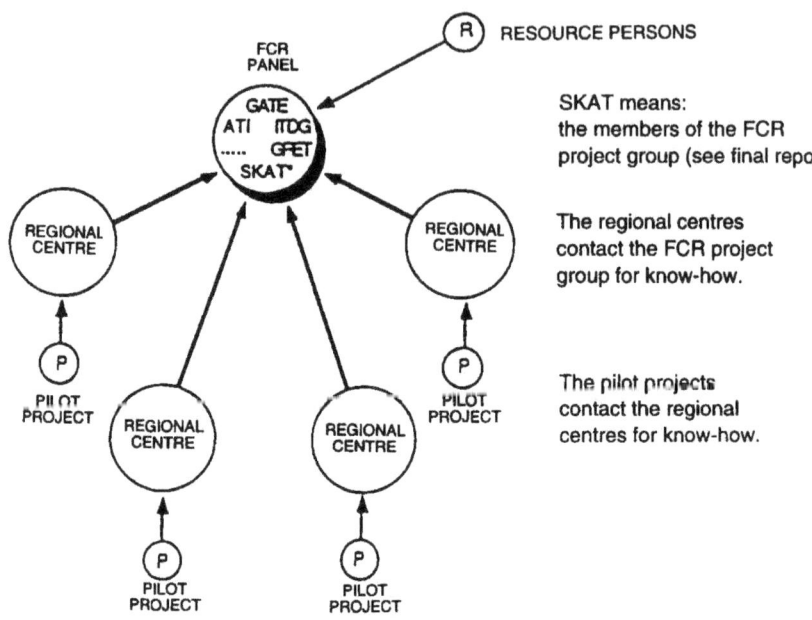

PHASE 2

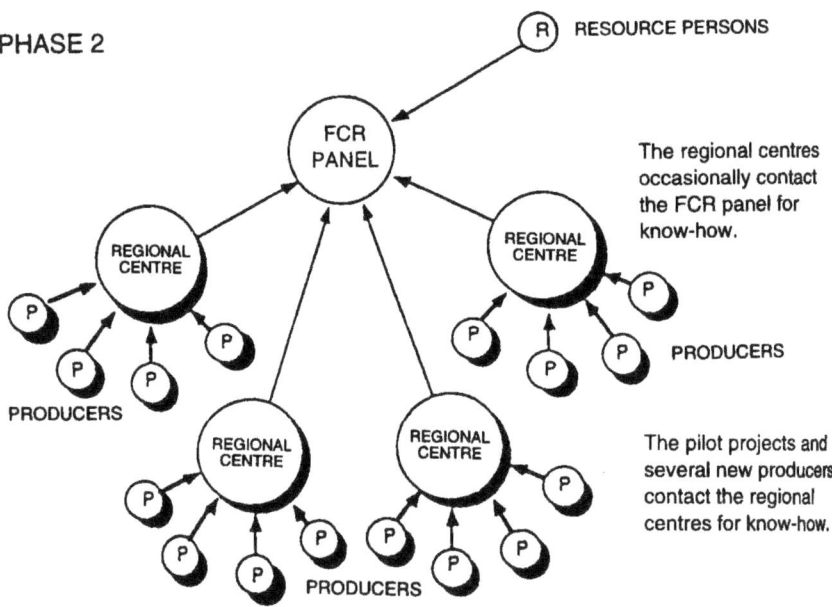

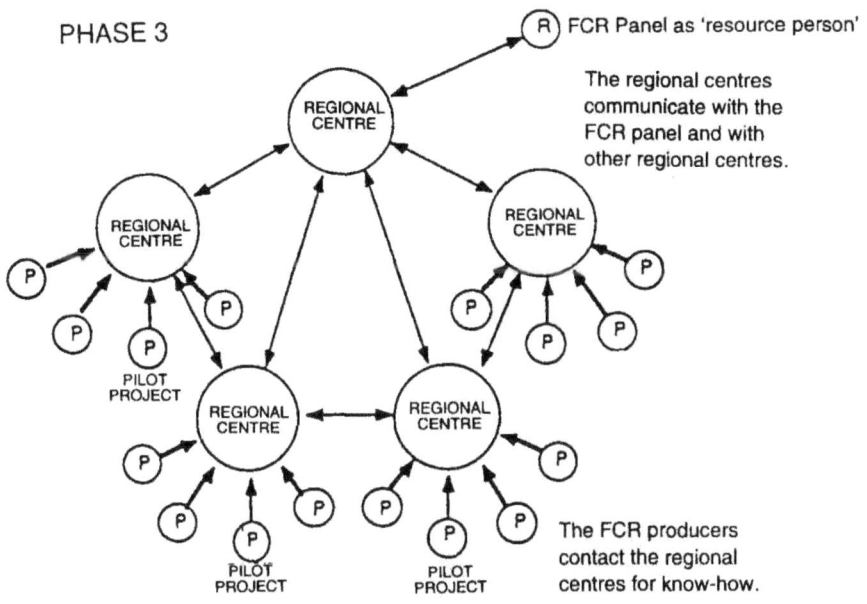

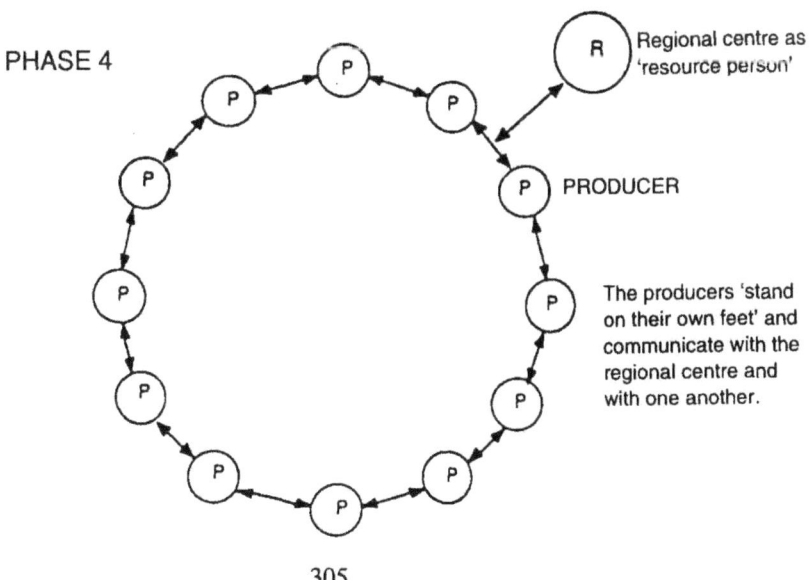

Fourth phase
The FCR technology has become a mature and self-reliant technology which is supported by a network of local producers of equipment. It works mainly on a commercial basis like other small industrial processes.

SKAT activities in the network

From 1987 to 1989, the Fibre Concrete Roofing Advisory Service (FAS) started to build up its activities at SKAT. The activities of FAS were:

1. General services: networking, contact with FAS panel members, contact with institutions and individuals, creation of regional centres, participation in seminars
2. Enquiry and answer service
3. Know-how collection, creating databank of institutions and individuals, producing FCR newsletter
4. Tools for management and administration
5. Tools for know-how transfer
6. Technical rationale.

The 750 working days invested in this activity were financed by the Swiss Development Corporation. In 1989 the pilot phase for the Building Advisory Service and Information Network (BASIN) started. The FCR/MCR network is fully integrated in it. The 145 days' work to create the BASIN network and lead it into a pilot phase were funded by GTZ.

For 1990–93, a new project phase, funded by SDC, is going on. The activities are:

1. General services
2. Know-how collection
3. Tools
4. Projects.

In this phase, the network does not only concentrate on FCR/MCR but is also expanding to include roofing in general.

National centres' activities

The national centres are the information and know-how source for FCR producers in the different countries. They are in contact with panel members and should also establish networking with the other national centres.

The main activities of the pilot centres within countries are:

- training
- monitoring of new production units (quality control)
- product promotion and marketing

- carpentry and tile laying
- research and roof design
- initiation or follow up of standards

These centres should have prospects for continuing work. Experience up to now shows that pilot centres are not financially self supporting. Income-creation through selling roofing has failed in some places because of direct competition with the promoted entrepreneurs. There is not one case known to us in which royalty on a tile has been paid to the centre. Another difficulty lies in the objective of supporting entrepreneurship and not just providing training. For this reason, several institutions are not suitable for becoming a national centre in the FCR/MCR network. Some centres generate income through selling tiles at promotional prices for public buildings (school houses, bus stops) and selling services in carpentry or roof design. Product promotion and marketing are very important activities for which it is quite difficult to find specialists in the various countries.

Experiences with the FCR/MCR network

After six years' work, we are now in phase 2 of the model. We do not know how and whether the pilot centres do perform and whether they will become national centres. It is still not clear whether there will be a need for them. We know that it is expensive to organize seminars because of intensive preparation work and high costs. In the beginning of the FCR networking, activities were concentrated on the FCR study. In 1989, there were already six activities funded by different organizations integrated into the network. These multi-dimensional activities running on different tracks create synergies which we consider very important for the effectiveness of the network.

To guarantee continuity, despite the different activities, only two persons should be in charge of the network management at the FAS panel.

Findings

The creation of a network needs long-term commitment, is time consuming and complex. A network can be very effective but its initial creation cannot be financially self supporting. Several phases in the development of a network have to create themselves and cannot be forced from outside.

References

Chapter 1

1. CAR 1991, *Building Materials for Development: Strengthening national building materials production capacity; Case Studies of Technology Transfer and Diffusion.*
2. Chindaprasirt, P, 1983, *Low Cost Cement for Rural Areas*, Faculty of Engineering, Khon Kaen University, Thailand.
3. COOPIBO, 1985, 'The Case of a lime-pozzolana industry in Rwanda: the PPCT', unpublished report, COOPIBO, Rwanda.
4. Cramer, S, 1979, *Gypsum Production on Maoi Island*, Institut fur Geologie, Berlin.
5. Cramer, S, 1979, *Gypsum Production; A Manual*, Institut fur Geologie, Berlin.
6. Government of Kerala, 1975, 'Report of the Committee constituted to study the question of making available lime and surkhi for building construction', unpublished report, India.
7. Kaplinsky, R, 1990, *The Economies of Small: Appropriate Technology in a Changing World*, IT Publications, London.
8. Kapur, PC and Sai, ASR, 1980, *ASHMENT Cement*, Indian Institute of Technology, Kanpur.
9. Li Mingyu et al. 1984, *Cement Industry in China*, Asian Pacific Region Mini-cement Workshop, UNIDO, Vienna.
10. National Council on Building Materials, 1990, 'Mini-cement in India', unpublished report.
11. Nohlier, M, 1986, *Construire en Platre*, Editions L'Harmatten, Paris.
12. Pagtambayayong, 1984, 'Progress Report on Pozzolanic Cement Binder Project', unpublished report.
13. Sakula J and Sauni, J, 1980, *Oldonyo Sambu Pozzolime Industry; History, Operation and Development*, SIDO, Tanzania.
14. Shrestha, R, 1985, 'Setswana Cement', unpublished report.
15. SIDO, 1981, *Evaluation report of Phase 1 lime-pozzolana Project at Oldonyo Sambu Ward, Arusha Region*, Small Industry Development Organization, Tanzania.
16. Sinha, S, 1990, *Mini-cement: a review of Indian experience*, IT Publications, London.
17. Smith, RG, 1984, *Rice Husk Ash Cement: Progress in development and application*, IT Publications, London.
18. Smith, RG, 1982, 'Small-scale production of gypsum plaster for building in the Cape Verde Islands', *Appropriate Technology*, IT Publications, London.
19. Spence, RJS, 1980, *Appropriate Technologies for Small-Scale Production of Cement and Cementitious Materials*, IT Publications, London.
20. UNCHS 1985, *The use of selected indigenous building materials with potential for wide application in developing countries*, UNCHS (Habitat), Nairobi.
21. UNCHS 1988, *A Compendium of Information on Selected Low Cost Building Materials*, UNCHS (Habitat), Nairobi.
22. UNIDO, 1986, *Lime in Development*, Vienna.

Chapter 4

1. The World Bank, May 1991, *World Development Report 1991*, Washington D.C.
2. IADB, 1991, *Economic and Social Progress in Latin America 1991 Report*. Inter-American Development Bank, Washington D.C.
3. Lola, C, 1988, *Lime Processing Project-Costa Rica Experimental Phase Final Technical Evaluation*, A.T. International, Washington D.C.
4. Lola, C, and Boynton, R, July 1991, 'An Overview of Small-Scale Lime Production'. *Proceedings of the Regional Workshop on Small-Scale Lime Production in Central America.* UNCHS-ATI-ITCR-FUNDATEC, Costa Rica.
5. Lola, C, October 1991, 'Regional Programme for Small-Scale Lime Production in Central America. Formulation Report', A.T. International, Washington D.C.
6. Rosales, E, July 1991, 'La Industria de la Cal en Costa Rica'. *Proceedings of the Regional Workshop on Small-Scale Lime Production in Central America.* UNCHS-ATI-ITCR-FUNDATEC, Costa Rica.
7. Quinonez, J, Robledo, C and Cuerra, E, July 1991, 'Perfil Descriptivo de la Industria de la Cal en Guatemala'. *Proceedings of the Regional Workshop on Small-Scale Lime Production in Central America.* UNCHS-ATI-ITCR-FUNDATEC, July 1991. Costa Rica.
8. Carpio, M, July 1991, 'Situacion Actual de la Industria de la Cal en El Salvador'. *Proceedings of the Regional Workshop on Small-Scale Lime Production in Central America.* UNCHS-ATI-ITCR-FUNDATEC, July 1991. Costa Rica.
9. Aguirre, E, July, 1991, 'Diagnostico de la Produccion de Cal en Nicaragua'. *Proceedings of the Regional Workshop on Small-Scale Lime Production in Central America.* UNCHS-ATI-ITCR-FUNDATEC, Costa Rica.
10. Beneditt, L, July 1991, 'Perfil Descriptivo de la Industria de Cal en Honduras'. *Proceedings of the Regional Workshop on Small-Scale Lime Production in Central America.* UNCHS-ATI-ITCR-FUNDATEC, Costa Rica.
11. Garcia, E, July 1991, 'The Lime Industry of Belize'. *Proceedings of the Regional Workshop on Small-Scale Lime Production in Central America.* UNCHS-ATI-ITCR-FUNDATEC, Costa Rica.
12. Jeans, A, Hyman, E and O'Donnell, M, 1990, 'Technology: The Key to Increasing the Productivity of Microenterprises'. Bethesda, MD: Development Alternatives, Inc. GEMINI Working Paper No. 8, Prepared for USAID.

Chapter 5

1. Malquori, G, 1960, 'Portland-pozzolana cement', Fourth International Symposium. Washington, Vol. 11, pp 983–1006.
2. Hammond, AA, 1986, 'Prospects and strategies for local building materials development in Africa'. Proceedings of CIB 86 Congress, Washington, Vol. 5, 1771–78.
3. Hammond, AA, 1963, 'Pozzolana cements for Low-Cost Housing', Proceedings of International Symposium on Appropriate Building Materials for Low-cost Housing. Vol. 1, CIB, UNCHS and RILEM, Nairobi, Kenya, pp 1–128.
4. Moavenzadeh, F, 1973, 'Transfer and adaptation of technology in the construction industry', Symposium on US Bilateral AID Strategies and programme in selected Areas of Science and Technology, Cornell University, USA.
5. UNECA, 1978, 'Components of the policy and strategy for the development of construction and building materials industries', E/CN.14/HUS/23, Addis Ababa, pp 1–14.

6. UNECA, 1983, 'Needs, constraints and prospects of African countries regarding availability of building materials', Proceedings, Inter. Symposium on Appropriate Building Materials for Low-Cost Housing Vol. 2, CIB, RILEM & UNCHS, Nairobi, pp 8–15.
7. UNECA, 1978, 'Construction and building materials industries in Africa', E/Cn.14/HUS/22, Addis Ababa, pp. 1–20.
8. UNCHS, 1987, 'Standards and Specifications for soil blocks, burnt clay bricks, lime, pozzolana and fibre-concrete roofing tiles', Background papers ARSO/CSC/HS/87/WK1/5. ARSO/CSC/UNCHS Workshop on formulation of Standards and Specifications for Local Building Materials, Nairobi.
9. UNCHS, 1985, 'The use of selected indigenous building materials with potential or wide application in developing countries', UNCHS (Habitat), Nairobi.
10. Hammond, AA 1979, 'Development of Building Materials Industries in Ghana', Conference of African Experts on Building Materials and Construction Industries in Africa. E.C.A., Addis Ababa.
11. Hammond, AA, 1974, 'Pozzolana from bauxite-waste', Conference on cost reduction in public construction sector, Accra, June 25–28 1974, B.R.R.I., Kumasi.
12. Anon, 1983, 'Republic of Botswana – Building Materials Sector Study', Ministry of Commerce and Industry, Gaborone, Republic of Botswana, pp 1–128.
13. United Nations, 1965, Official Records of the General Assembly Twentieth Session Supplement No. 14 (A/6014), Resolution 2036 (XX) adopted 7th December 1965, page. 39, para (c).

Chapter 6

Mamu, RR and Hill, NR, 1982, 'Potential pozzolanic material and raw materials for lime making in Karonga District'. Report RRM26/NRH2, Geological Survey of Malawi.

Chapter 7

1. Herath, JW, 1975, 'Mineral Resources of Sri Lanka', Economic Bulletin, No. 2, Geological Survey, Sri Lanka, 1975, 57.
2. Herath, JW, 1980, 'Mineral Resources of Sri Lanka', 2nd Revised Edition, Economic Bulletin No. 2, Geological Survey Department, Sri Lanka, 29. (Sinhalese edition available).
3. Herath, JW, 1985, 'The Economic Geology of Sri Lanka', *Natural Resources Series* – No. I, NARESA Publication, 155.
4. Herath, JW, 1990, 'Economic Geology – Sri Lanka Minerals: A descriptive catalogue'. CRDC TIS.

Chapter 8

1. Allen, WJ, 1985, 'Field Investigation of Pozzolanic Materials, Koru, Kenya', Intermediate Technology Development Group.
2. Allen, WJ and Spence, RJS, 1980, 'The properties and testing of lime-pozzolana mixtures', Building Research Worldwide, Eighth CIB Triannial Congress, Oslo, June 1980.
3. Allen WJ, 1981, 'The evaluation and testing of a volcanic pozzolana', PhD Thesis, University of Cambridge, July 1981.

Chapter 9

1. Hammond, AA, 1990, 'Research on Rice Husk Ash binders low-cost housing technology'. Prepared for GTZ. HRDU, University of Nairobi, p 79.
2. Kamau, GN, 1991, 'Chemical and Mineralogical Analysis of Rice Husk Ash'. Unpublished report, Nairobi.
3. Mbindyo, KN, Kamau, GN and Tuts, R, 1991, 'Recycling of rice husk wastes for use as a cement replacement material in Kenya'. Paper at the 'Environment and Development' session of the world conference on Philosophy, Nairobi, 1991, p 14.
4. Tuts, R, 1990, 'Pre-feasibility study on the use of Rice Husk Ash as cementitious binder in Kenya'. HRDU, University of Nairobi, p 71.
5. Wahome, ER, 'Properties of local pozzolanic materials for use in concrete'. MSc, thesis, University of Nairobi, 1990, p 256.
6. ARSO/CSC/UNCHS, *Standards and specifications for local building materials; Technical Papers presented at the ARSO/CSC/UNCHS Workshop.* Nairobi, 1987, p 110.
7. Bureau of Indian Standards, *Specifications for lime-pozzolana mixture (IS:4098–1983).* New Delhi, reprint 1989, p 9.
8. Bureau of Indian Standards, *Specifications for pozzolana-portland cement (IS:1489–1976).* New Delhi, reprint 1989, p 12.
9. CBS, 1988, *Statistical Abstract 1988.* Central Bureau of Statistics.
10. Chopra, SK, 1979, 'Utilization of rice husk for making cement and cement like binders', In: *Proceedings of joint workshop on Production of cement-like materials from agro-wastes.* Bangalore, pp 135–49.
11. Dass, A and Rai, M, 1979, 'Prospects and problems in the production of cementitious materials from rice-husk'. *Proceedings of joint workshop on Production of cement-like materials from agro-wastes.* Bangalore, pp 49–56.
12. Davis, Benfield and Everest, 1988, *Spon's International Construction Costs Handbook.* London, p 369.
13. Dramais, J, 1986, *Etude des conditions technico-économiques permettant le développement d'une production de liants pouzzolaniques au Kenya.* Paris, p 27.
14. Republic of Kenya, 1990, *Restrictive trade practices, monopolies and price controls act: Cement price list.* Nairobi, p 11.
15. Shrestha, RD, 1979, 'Industrial Utilization of Agro-Wastes for Making Cement-Like Materials'. *Proceedings of joint workshop on Production of cement-like materials from agro-wastes.* Bangalore, pp 77–108.
16. Smith, RG, 1984, *Rice Husk Ash, Progress in Development and Application.* IT Publications, p 45.
17. Spence, RJS, 1980, *Small scale production of cementitious materials.* London, 1980, p 49.
18. Spiropoulos, J, 1985, *Small Scale Production of Lime for Building.* GATE, pp 11–24.
19. Tuts, R, Verschure, H and Wouters, L, 1987, *Reader: a compilation of texts related to the PPCT Project, Rwanda.* PGCHS – Coopibo, p 86.
20. Tuts, R, 1990, *Construction Costs in Kenya, 1980–1989: an analytical overview of costs, indices and weights.* HRDU, p 56.
21. UNCHS, 1987, Standards and Specifications for soil blocks, burnt-clay bricks, lime, pozzolanas and fibre concrete roofing, in *Workshop on formulation of standards and specifications for local building materials.* Nairobi, p 55.
22. UNCHS and CSC, 1989, *Journal of the network of African countries on local building materials and technologies.* Vol. 1, No. 1, p 21.

Newspaper cuttings

23. Nduati, S, 'Kenya to import Egypt's cement'. In: *Daily Nation*, Nairobi, 9 Feb. 1990.
24. Kamanga, D, 'Cementing Kenya's French Connection'. In: *Sunday Times*, Nairobi, 11 Feb. 1990.
25. Mwangi, P, 'Kenya may have to import cement' – report. In: *The Standard*, Nairobi, 11 Apr. 1991.
26. Kencem, 'Recommended retail prices of cement'. In: *The Standard*, Nairobi, 23 May 1991.
27. Chege Wa Gachamba, 'Giant rice scheme to boost production'. In: *Daily Nation*, Nairobi, 20 June 1991.

Chapter 10

1. Agrostat database, 1991, 'Paddy rice production from 1961 to 1990', Statistics Division, Food and Agriculture Organization of the United Nations, Rome.
2. Hough, JH and Barr, IIT, 1956, 'Possible uses for waste rice husks in building materials and other products', Bulletin No. 507, Agricultural Experimental Station, Louisiana State University, USA.
3. Mehta, PK, 1975, 'Rice husk ash cement – High quality, acid resisting', *Journal of the American Concrete Institute*, 72, (5), pp 235–6, ACI, New York.
4. RCTT, 1979, 'Rice husk ash cement', Proceedings of a joint workshop, Peshawar, Pakistan, January 1979. Rural Central for Technology Transfer, Bangalore, India.
5. Smith, RG, 1984, *Rice husk ash cement: Progress on development and application*, 45 pp, Intermediate Technology Publications, London, England.
6. Smith, RG, 1984, 'Rice husk ash cement: Small-scale production for low cost housing', Proceedings of international conference on low cost housing for developing countries, CBRI, Roorkee, India, pp 687–95. Sarita Prakashan, Meerut, India.
7. Cook, DJ, 1985, *Rice husk ash cements: Their development and applications*, United Nations Industrial Development Organization, Vienna, Austria.
8. British Standards Institution, 1970, 'Methods of testing cement, BS 4550:1970', B.S.I., London, England.
9. Indian Standards Institution, 1976, 'Specification for lime-pozzolana mixture, IS 4098', I.S.I., New Delhi, India.
10. Smith, RG, 1988, Consultant's report on production of rice husk ash cement in Suriname, 19 pp, UNCHS (HABITAT), Nairobi, Kenya.
11. Kapur, PC, 1981, 'TiB: tube in basket rice husk burner for producing energy and reactive rice husk ash: Proceedings of third workshop on rice husk ash cement, New Delhi, India, Rural Centre for Technology Transfer, Bangalore, India.
12. Smith, RG and Kamwanja, GA, 1986, 'The use of rice husks for making a cementitious material.' Proceedings of a CIB/RILEM international symposium on use of vegetable plants and their fibres as building materials, pp E 85–94, October 1986, National Centre for Construction Laboratories, Baghdad, Iraq.
13. Smith, RG and Tait, G, 1989, 'Rice husk ash cement in Guyana.' Proceedings of the Third CIB/RILEM symposium on materials and technologies for construction of low-cost housing, Mexico City, November 1989, pp 229–38, Instituto del Fondo Nacional de la Vivienda para los Trabajadores, Mexico.

Chapter 11

Perry and Chilton Chemical Engineers Handbook Fifth Edition, 1973, McGraw Hill.

Boynton, Robert S, *Chemistry and Technology of Lime and Limestone*.

1989 Annual Book of ASTM Standards Section 4, Volume 04.02: Concrete and aggregates.

Prasher, CL, 1987, *Crushing and Grinding Handbook*.

Taggart. Arthur F, *Handbook of Mineral Dressing Ores and Industrial Minerals*. Arthur F Taggart.

'Symposium of Use of Pozzolanic Materials in Mortars and Concretes'. Presented at the First Pacific area National Meeting – American Society for Testing Materials, San Francisco, California Oct. 10th to 14th 1949.

The Swedish Cement Association, 'Proceedings of the Symposium on the Chemistry of Cements Stockholm 1938'. Held under the Auspices of the Royal Swedish Institute for Engineering Research.

Chemistry of Cement – Proceedings of the Fourth International Symposium, Washington 1960. Volume 11. National Bureau of Standards Monograph 43.

'Portland Pozzolan Cement – Working Report No. 31'. Mtui, AL and Kawiche, CM, 1983, National Housing and Building Research Unit, Dar-es-Salaam.

Chapter 12

1. Koss, Pacheco, Rosales, *Informes Finales, Primera Fase*, Proyecto Hornos de Cal. ITCR-Fundatec-ATI, 1988.
2. Technoserve, 'Estudio de mercado de la cal en Costa Rica', 1988.
3. Tanaka, A, *Análisis Comercial del Proyecto Hornos de Cal*.

Chapter 15

1. Boynton, RS, 1980, *Chemistry and Technology of Lime and Limestone*, John Wiley and Sons, Inc. New York.
2. Bush, A, 1989, 'Balaka Lime Kiln: Report of trials conducted in April 1989'. ITDG report.
3. Jones, B, (Aptech Zimbabwe), 1990, 'Balaka Lime Project: Final installation and commissioning report'. ITDG report.
4. Spiropoulos, J, 1987, 'Balaka Lime Project: Report and draft project proposal'.
5. Wingate, M, 1988, *Small scale lime burning: A practical introduction.* IT Publications, London.
6. Spiropoulos, J, 1992, *Chenkumbi Lime: fuel-efficient with quality production.* IT Publications, London.

Chapter 19

1. Ahmad, Syed Faiz, 1989, 'A comparison of Pozzolanic materials, Neo and Classical', *Proc. of the 2nd EASEC Conference*, Chiang Mai, Thailand.
2. Ranasinghe, AP, 1985, 'Use of Rice Straw Ash as Pozzolana', *M. Engg. Thesis No. St–85–1*, Asian Institute of Technology, Bangkok, Thailand.
3. Neville, AM, 1978, *Properties of Concrete, 2nd ed., ELBS, Pitman Publishing*, London.

Chapter 21

1. 'Rapport sur le développement des constructions dans le cadre de la valorisation des matériaux locaux'. Rapport interministériel, Alger, Sept 1984.
2. 'Carte nationale des substances utiles pour les matériaux de construction'. Agence Nationale pour l'Aménagement du Territoire. Alger 1987.
3. Moussa, J, 'Phase bibliographie' de l'étude sur les produits silico-calcaires', Rapport interne, CNERIB, Décembre 1989.
4. *La chaux: sa production et son utilisation dans l'habitat.* Groupe de Recherche et d'Echanges Technologiques.
5. La chaux: ses utilisations. Ed. Balthazard, Cotte & Techno-Nathan 1990.
6. *La chaux: pour batir et décorer*: Balthazard, Cotte & Nathan pratique 1990.
7. Debonviére, J, 1983, La chaux dans les pays en développement: une alternative au ciment?

Chapter 22

1. Mehra, SR, 1965, 'A Note on Manufacture of packaged Lime-reactive Surkhi Mixture for Use in Building Mortars and Foundation Concrete', issued by Central Road Research Institute, New Delhi.
2. CRRI, 1964, 'Pozzolanic Clays of India, their Industrial Exploitation and use in Engineering Works', Central Road Research Institute, New Delhi, Special Report No. 1.
3. Rao, AVR, 1970, 'Fluidized Bed Calciner for the Heat Action of the Clays', *NBO Journal*, 15(1) 1–2.
4. *Perry's Chemical Engineering Handbook*, 3rd edn., 1950, McGraw-Hill Book Co., London.
5. I.S. Specification No. 1344–1968.
6. Indian Patent No.131628/71.
7. Sen Gupta, J and Rao, AVR, 1978, 'Production of Clay Pozzolana by Fluidized Bed Technique', Trans of Ceramic Society. Vo–XXXVIII. No–4.
8. Swamy, RN, 'Cement Replacement Material'.
9. Hammond, AA, 1983, 'Pozzolana Cement for Low-Cost Housing', Proceedings of Symposium on Appropriate Building Materials for Low-Cost Housing.
10. Dave, NG, Malhotra, SK and Verma, ML, 1985, 'Energy Advantages of Lime-based Mortars', *NBO Journal*, Vol. XXX. No. 2.

Chapter 23

1. Ashhurst, John, 1983, *Mortars, plasters and renders in conservation*, EASA, London.
2. Oliveira, Mário, d'Affonsêca, Silvia and Santiago, Cybèle, 1990, The study of accelerated carbonation of lime-stabilized soils. *Adobe 90 Preprints. 6th International Conference on the Conservation of Earthern Architecture*, 166–170.
3. Hodgson, Fred, 1920, Cyclopedia of bricklaying, stone, masonry, concretes, stuccos and plasters. Chicago, Frederick Drake.
4. Anton, Maria Ruiz de, 1987, Historical exterior surface treatments – architectural renderings. New York, unpublished dissertation.
5. Santiago, Cybèle Celestino, 1991, Organic additives in lime-mortars. Salvador, unpublished dissertation.
6. Plinius, C Secundus, 1962, *Natural history* (transl. by Eichholz, DE) London.
7. Ellis, Myriam, 1966, As feitorias baleeiras meridionais do Brazil. Sâo Paulo, unpublished dissertation.

8. Sickels, Lauren-Brook, 1981, Organic additives in mortars. *Ear*, 8, 7–20.
9. Santiago, Cybèle Celestino, 1991, Organic additives in lime-mortars. Salvador, unpublished dissertation.
10. Phillips, Morgan, 1974, SPNEA–APT Conference on mortars. *APT Bulletin*, 6, 1, 9–39.
11. Rodrigues, Luiz Erlon, 1984, Alterações bioquímicas lisossômicas nas esquistossomose mansônica hepática experimental. Salvador, unpublished dissertation.

Chapter 24

1. Meyers, JF, Pichumani, R and Kapples, BS, 1976, 'Fly Ash–A Highway Construction Material', USDOT, FHWA–IP–76–16.
2. Usmen, MA, Bowders, JJ, 1990, 'Stabilization Characteristics of Class F Fly Ash'. Paper accepted by Publication by the Transportation Research Board, Washington DC.
3. Glen, F, 1987, 'Recycling Pavements with Self Cementing Fly Ash', EPRI Proceedings: Eight International Ash Utilization Symposium, Vol. 2. pp s12-1–12.
4. Kraft, DC, et al, 1979, 'Ash Utilization in Bikeway Construction'. Proceedings: Fifth International Ash Utilization Symposium, METC/SP–79/110 (Pt.2), pp 694–712.
5. Sutherland, HB, Finlay, TW and Cram, IA, 1968, 'Engineering and Related Properties of Pulverized Fuel Ash', *Journal of the Institution of Highway Engineers*, Vol. 15, pp 19–35.
6. Gray, DH, and Lin, YK, 1972, 'Engineering Properties of Compacted Fly Ash', *Journal of the Soil Mechanics and Foundations Division*, ADCE, Vol. 98, No. SM4, pp 361–380.
7. Bowders, JJ, Gidley, JS and Usmen, MA, 1990, 'Permeability and Leachate characteristics of Class F Fly Ash', paper accepted for publication by the Transportation Research Board, Washington DC.
8. Baradan, B, Usmen, M and Yazici, S, 1992, 'Engineering Properties of Lime and Cement Stabilized Lignite Fly Ash', American Coal Ash Association 9th International Coal Ash Symposium, Orlando, Florida, pp 43-1–17.

Chapter 25

1. Janice de Area Leao Schilderman, 1985, 'Vers l'utilisation du liant pouzzolanique dans l'habitat informel', COOPIBO, Leuven.
2. Schilderman, Theo, 1986, 'Le liant pouzzolanique LIPORWA: Caracteristiques et emploi', PPCT, Ruhengeri.
3. Schilderman, Theo, 1987, 'Het PPCT in Rwanda: wel en wee van een NGO intervetie in de kleinschalige industrie', in BOW Nieuwsbrief 41, Wageningen.
4. Schilderman, Theo and Vershure, Han, 1988, 'A local alternative to Portland Cement in Rwanda', in Bertha Turner, *Building Community*, London.
5. Snelder, Herman, 1989, 'Project Pouzzolanes Chaux Tourbe: Beschrijving en analyse van een klein industrieel projekt in Rwanda (1978–88)', University of Twente, Netherlands.
6. Gahunde, Christophe, 1991, 'Project Pouzzolanes Chaux Tourbe: Rapport d'Activities 1990', Ruhengeri.

Chapter 26

1. Redgrave, Gilbert R and Spackman, Charles, 1924, *Calcareous cements*. Charles Griffin & Co. Ltd, London.
2. Cowper AD, 1927, *Lime and lime mortars*. Department of Scientific and Industrial Research Building Research Special Report No. 9 HMSO, London.
3. Gunawardhana HD and Dias PPSP, 1987, 'Use of dolomite for finishing coat in masonry'. *J. Natn. Sci. Coun., Sri Lanka*, 15(1): 1–10.
4. Vasari Georgio, tr. Maclehose Louisa, 1960, *Vasari on Technique*. First Published 1550. Translation published 1907. Reprinted Dover Publications inc, New York.
5. Vicat LJ, tr. Smith J, 1837, *Treatise on Calcareous Mortars*. London.

Chapter 27

1. Holmstrom, I, 1977, 'Suitable materials for use in repair of historic structures', Conference on structural conservation of historic buildings B–19/9/1977 Rome (a paper).
2. Litvan, G, 1986, 'Experiments with particulate admixtures to concrete' (a paper).

Chapter 29

1. Markus, TA, 1955, 'Design Techniques for Earth Housing'. MSc Thesis, Massachusetts Institute of Technology.
2. Volunteers in Technical Assistance, 1977, 'Making building blocks with CINVA-Ram block press'. M Raine. Maryland USA.
3. Clare, KE and Cruchley, AE, 1957, 'Laboratory Experiments in the Stabilization of Clays with Hydrated Lime', Geotechnique 7.
4. UNCHS, 1990, 'Co-operation in The African Region on Technologies and Standards for Local Building Materials'. UNCHS (Habitat) Nairobi.

Chapter 32

'The Landmark Handbook 1991', obtainable from the Landmark Trust, Shottesbrooke, Maidenhead, Berkshire, SL6 3SW, UK.

Chapter 33

1. Dave, NG and Masood, I, 1972, 'Studies on the existing lime kilns in India' Seminar on Lime: Manufacture and uses, New Delhi.
2. Srinivasan, NR, 1956, 'Surkhi (burnt clay) as a pozzolana' Road Research Paper No. 1, CRRI, New Delhi.
3. Ghosh, RK, Srinivasan, NR, Krishnamachari, R and Bhaki, KL, 1966, 'Development of Clay Pozzolana Industry, Pt. II – Manufacture of Reactive Surkhi Mixture' *Research and Industry*, 3(2), pp 164–169.
4. Sen Gupta, I and Rao, AVR, 1977, 'Production of clay pozzolana by Fluidized bed technique' Consultation on lime pozzolana, New Delhi.
5. Thatte, CD and Patel, JK, 1977, 'Production of pozzolana in vertical shaft kilns. Consultation on lime pozzolana, New Delhi.
6. Rajinder Kumar and Khazanchi, AC, 1988, 'Utilization of flyash – A new approach' National workshop on utilization of fly ash, Roorkee.

7. Datta, RK and Dass, R, 1979, 'Development of reactive pozzolana from rice husk and clay' UNIDO/ESCAP/RCTI Workshop on Utilization of rice husk for making cement like building materials, Peshawar, Pakistan, 1979.
8. Chopra, SK, 1981, 'Development, production and use of rice husk ash cement in India', RCTT workshop on rice husk ash cement, New Delhi.
9. Das, A and Malhotra, SK, 1985, 'Cementitious binder from waste lime and rice husk', Project proposal No. 59, CBRI, Roorkee.
10. IS: 2250-1965, 'Code of Practice for preparation and use of masonry mortars', Bureau of Indian Standards, New Delhi (India).
11. Masood, I, Dave, NG, Mehrotra, SP, Malhotra, SK & Verma, MK, 1983, 'Activated Lime Pozzolana Mixture (ALPM): A promising cementitious binder for developing countries', International Conference on materials, construction techniques, elements appropriate for economical housing in developing countries, Paris, Jan. 25-27.
12. Chopra, SK, Ahluwalia, SC, Laxmi, S, Ali, MM and Gopal, S, 1981, 'Technology and manufacture of rice husk ash masonry cement', RCTT workshop on rice husk ash cement, New Delhi.
13. Dave, NG, Hazela, RN, Verma, CL, Aslam, M and Agarwal, AL, 1987, 'Building materials industries and pollution', National Workshop on Building materials Technologies and Pollution abatement, New Delhi.
15. Dass, A and Malhotra, SK, 1979, 'Lime stabilized soil as a building material – Preliminary investigations', *NBO Journal* 24(2), pp 29-30.
16. Phull, YR and Sethi, KL, 1977, 'Role of Lime flyash mixture in road construction', Consultation on lime pozzolana, New Delhi.
17. Chopra, SK, Taneja, CA and Tehri, SP, 1968, 'Development of Cellular Concrete based on lime and flyash', Research and Industry, 13(4).
18. *Calcium silicate bricks.* Special publication, CBRI, Roorkee.
19. Dave, NG, 1977, 'Energy savings on lime pozzolana', consultation on lime pozzolana, New Delhi.
20. Dave, NG, Malhotra, SK and Verma, ML, 1985, 'Energy Advantages of lime based materials' *NBO Journal*, Vol. 30.
21. Dave, NG, Verma, CL, Mehrotra, SP, and Verma, ML, 1983, 'Investigation of kankar burning processes', Khadigramodyog, pp 181-186.
22. Annual Report, 1990, 'Khadi and Village Industries Commission', Bombay.
23. Khonijo, MK, Ahuja, YL and Rao, AVR, 1977, 'Employment potential of the Lime and Pozzolana Industry', Consultation on Lime Pozzolana, New Delhi.

Appendix 1b

The recipes from Venice are from: The European Centre for Craftsmen, Isola di San Servolo, Venice.
The recipes from Rome are from: The ICCROM Workshops, via Del Porto, Rome.
The ancient Roman stucco is taken from: The Ten Books on Architecture, by Marcus Vitruvius Pollio (book VII, chapter III), circa 16 BC.

Appendix 3

1. Gram, HE, Parry, JPM, Rhyner, K, Schaffner, B, Stulz, R, Wehrle, K and Wehrli, H, 1989, *Fibre Concrete Roofing.* SKAT/IT Publications Ltd.

www.ingramcontent.com/pod-product-compliance
Ingram Content Group UK Ltd.
Pitfield, Milton Keynes, MK11 3LW, UK
UKHW021844140426
5217IPUK00022B/1575